The Spirit of
Dan-Air
Services
- 40 Glorious years!

Graham M. Simons

THE SPIRIT OF DAN-AIR

First published 1993
by GMS Enterprises
67 Pyhill, Bretton
Peterborough

© Graham M Simons

ISBN 1 870384 20 2

AUTHORS ACKNOWLEDGEMENTS

This book has been written with the co-operation of so many people, it is impossible to list them all. Nevertheless, without their help and support, the book would never have got into print.
I thank you all.

Although possibly unfair to single out anyone, there are a number of people who, through their dedication 'above and beyond the call of duty' deserve my specific thanks:

John Mayes for picking up the pieces of the book following the take-over, for guiding me unerringly through Dan-Air and for providing contact with so many people so late in the day. **Mr F. E. F. Newman** for providing an insight into 37 years. **Captain Arthur Larkman** for providing so much insight into the operations of the Company, especially the earlier days. **Captains Burgess, Overbury, Smith** and **Sintes** for letting me have an invaluable insight into their careers. **Keith Moody** for insight, help and encouragement. **Lindy Simmons** for providing so many first names to initials. **Margret Gray** for chatting so freely in Mahon despite being so busy. **Peter Carter** for freely providing so much information. **David James** for taking time to explain the background of so much. **Mike Forsyth** for his help on Ops and so many photographs. **Mo Perera**, who will forever be my favorite stewardess. **Frankie Ross** and **Margaret Barton** for fielding my frantic telephone calls. **Rosemary** & **Roger Cooper**, for being so dedicated. **Alison Beedie, Richard Foot, John Hamlin, Suzi, Cap'n Mike, Peter and Trish, Ed, Andy, Pete S-H.**
and finally...

my long-suffering wife Anne

The main text of this book
has been set in 12pt 'Times'
on an Apple Macintosh DTP system

All rights reserved. No part of this publication may be reproduced, stored in a retrieval system, or transmitted, in any form or by any means, electronic, mechanical, photocopying, recording or otherwise without the prior permission of the publishers.

Printed and bound by
Woolnough Ltd
Express Works
Irthlingborough
Northants

THE SPIRIT OF DAN-AIR

CONTENTS

	Foreword	5
	Introduction	7
Chapter 1	*Birth of an airline*	9
Chapter 2	*The first move*	15
Chapter 3	*The move to Gatwick*	31
Chapter 4	*Comet!*	47
Chapter 5	*Modernisation*	63
Chapter 6	*Pax, Cargo, Oil and the 748*	93
Chapter 7	*Flying equals*	109
Chapter 8	*New aircraft and new routes*	119
Chapter 9	*A new generation - a new service!*	145
Chapter 10	*To the peak - and beyond*	173
Chapter 11	*So near disaster*	179
Chapter 12	*The news breaks*	199
Chapter 13	*A personal view*	213
Appendix 1	*The Directors*	223
Appendix 2	*The fleet*	225
Appendix 3	*Key Dates*	241

THE SPIRIT OF DAN-AIR

*This book is dedicated
to every member of staff of
Dan-Air Services Ltd
and all its
associated Companies*

THE SPIRIT OF DAN-AIR

FOREWORD
by
Mr. F. E. F. Newman M.C., C.B.E

I am very pleased to have been given the opportunity to write a Foreword to this book about Dan-Air, which covers 40 years of aviation endeavour. Graham Simons has worked extremely hard to uncover a wealth of information about it's people, aircraft and commercial activities over these years and has truly produced a book of great interest to its readers.

Right from the early days, when Dan-Air was in its infancy, through the middle years striving successfully for a substantial share of the package holiday business, to the final days when fighting for survival at a time of recession, the efforts of every member of the staff was always positive and contributed to the successes achieved.

I believe that when we used the name 'Dan-Air Services', we intended the word 'Services' to have two meanings. The first implying an air service and the second an implication of giving a 'service' to our customers. In spite of the many difficulties, our staff have always put this second objective first and their activities will be remembered throughout the United Kingdom and Europe by our passengers and Tour Operators alike.

I am grateful to Graham Simons for all he has done to produce this record, also, to our loyal staff, and I hope that they will look back with satisfaction to the years spent with Dan-Air.

THE SPIRIT OF DAN-AIR

THE SPIRIT OF DAN-AIR

INTRODUCTION

The overall saga of the United Kingdom's highly competitive civil airline network since the end of World War Two is littered with the names of aerial enterprises that bloomed briefly, then disappeared - British Eagle, Channel, Court Line, Invicta, Laker, British Caledonian and, of late, Air Europe. This is the story of the longest serving independant airline still operating up to November 1992 under its original name and flight-code - Dan Air Services Limited.

From its humble beginnings as a subsidiary of the London-based ship brokers Davies and Newman in the early 1950s, Dan-Air steadily grew into what was one of the country's leading scheduled service operators that still retained a major presence in the charter field. For many years, until the privatisation of the state airlines, Dan-Air was the United Kingdom's largest independent airline - not only well known for its wide-ranging Inclusive Tour operations, but for an ever-expanding involvement in other spheres of air transport, lately making use of the most modern of equipment.

I knew that over the years Dan-Air had been involved in a number of events significant to the history of British civil aviation, and I suspected that there was much more than met the eye, for Dan-Air had never been one to sing its own praises. Since 1989 I had been preparing the background for this story; frankly, the potential of 'The story of Dan-Air' facinated me. So, when back in the autumn of 1991 I became involved in negotiations that eventually retained me to research, compile and publish the official history of Dan-Air, I - a self-confessed civil aviation 'nut' - was like a kid let loose in a toystore.

There I was, with full access to the United Kingdoms second largest airline. I could go where I wanted, speak to whoever I wanted and ask any questions I wanted; I was in paradise!

Despite my excitement for the forthcoming project, I was determined from the onset that this would be an historically accurate document. I would stay neutral, be just an observer and become a 'fly on the wall'. That attitude lasted less than twenty-four hours.

My 'letter of introduction', signed on the 30th January 1992 by Charles Whyte, the Group Managing Director no less, introduced me *"To whom it may concern'*, went on to state my task and then mentioned that *"This will require quite a considerable amount of research and it is hoped that you will provide any assistance necessary"*. That letter, although intended as an aid to 'smooth my passage' through the Company, was a complete waste of paper - I never had to use it once!

THE SPIRIT OF DAN-AIR

The reason I just could not stay neutral, nor ever had the need to use the letter can be summed up in one simple phrase - the people that made up the company. During 1992 they had their backs against the wall. The media were hounding them as to what was going on and the 'press' was 'bad', going on and on about a 'poor brand image' but never once did a single member of staff - whereever they were - fail to give the customers not only the service required, but also the personal service they deserved.

Despite the momentous events that occurred on 23rd October 1992, never before have I come across such a highly professional, enthusiastic, dedicated, friendly, helpful, loyal, happy and just plain nice bunch of men and women who genuinely wanted the full story to be told. How did they keep going? Over the years the staff had became used to those who tried to put the airline down. As Rosemary Cooper once summed it up for everyone, "*You've got to have faith*". No company could have wished for a better set of ambassadors.

In researching this book I had an almost unique 'overview' of the company and this, I quickly realised, was predominantly a 'people' story. Without them, going about their everyday jobs, there could not have been an airline.

And so the document evolved. It is as historically accurate as I can make it, given the passage of 40 years. Most of the people featured in the story should to be regarded as representatives for the many hundreds of others that were employed in similar positions; all have interesting tales to tell but it was not possible to include them all!

No doubt there are also omissions - it is impossible to include every last detail and keep the book interesting. To the casual reader, many of the events described herein may appear to be 'one-offs'; in many cases this is not so, they are just selected examples! There may also be errors - I hope they are few, and blame for all can all be laid at my feet.

It is my greatest hope that this book does justice to the story of one of the worlds great airlines and captures some of the wonderful feelings that pervaded every aspect of the operation; feelings that someone once termed '*The Spirit of Dan-Air*'. Every member of staff really did form part of one big happy family.

Graham M Simons

Peterborough
0658hrs 5 March 1993

DAN-AIR SERVICES

CHAPTER ONE
Birth of an airline...

The shipbroking company of Davies and Newman Limited was established in the City of London in 1922. The Company, with offices on the second floor of 3 Gracechurch Street, London E.C.3 specialised in the chartering of oil tankers, but was involved in other aspects of the shipping industry.

The Baltic Mercantile & Shipping Exchange Ltd, of which the Company was a member, decided in the late 1940s to open an airbroking section and Davies and Newman employed Mr Selby Lownes in his capacity as airbroker.

The Company were agents on the Exchange for Ciro's Aviation Ltd, the Gatwick-based operators of a pair of Douglas DC-3s and a pair of De Havilland DH 89 Dragon Rapides used for passenger charter and freight work. At this time air business was difficult and hard to find, so Davies and Newman also took an Agency for Meredith Air Transport, a small Southend-based *ad hoc* charter airline.

Meredith Air Transport had been formed by Captain W. E. Hamilton DFC DFM and registered on 12 August 1952 as the U.K. branch of a South African company called Tropic Airways. Office accommodation had already been rented by Mr Hamilton in London from Davies & Newman, so it was natural that they became Meredith's brokers and also offered financial help to the struggling Company in the form of a debenture on G-AMSU the single DC-3, they were operating. At Southend, Meredith Air Transport was ran by just Captain Hamilton and one secretary, and they were responsible for everything from cleaning and loading the aircraft to keeping the paperwork straight!

During December the DC-3 set off from Southend for a 14-day tour of the eastern Mediterranean, but on Christmas Day the machine over-ran the runway at Jerusalem, resulting in damage to the tailwheel. Replacement parts were flown out, enabling the aircraft to limp back to Southend for full repairs.

Despite making a number of aerial cruises around the Mediterranean in the early part of 1953, the Company's debts mounted and Meredith was forced to part with the aircraft.

It was a logically progressive step therefore, that when Meredith's found themselves in financial difficulties and the debenture was due for repayment, Davies and Newman should take over the aircraft and form their own airline, retaining Meredith to operate the aircraft under a six month contract. At the same time as gaining the machine, Davies & Newman also purchased a contract from them to operate a series of flights into Berlin.

DAN-AIR SERVICES

Take one aircraft...
With this single DC-3, Davies & Newman was determined to set up an air service, if not an airline, so Dan-Air Services Limited - the name derived by the simple expedient of taking the first letters of the names of the parent Company - was registered with capital of £5,000 as a subsidiary to Davies & Newman on 21 May 1953. An Aircraft Operators Certificate (AOC) was granted to the new Company and Meredith's old main base at Southend Airport was taken over. The new Company's directors were listed as J.W. Davies, F.E.F. Newman and E.O. Wallis with Captain F. R. Garside as the Chief Pilot.

The Douglas Dakota (or to give it the correct civil designation Douglas DC-3) was and is an all-metal, low-wing military derivative of this American manufacturer's famous design, with its roots firmly based in the 10-seat DC-1 design of 1933 powered by a pair of 875 horsepower Wright Cyclone piston engines. From this evolved the 14-seat DC-2 of 1935, followed by the 21-seat DC-3 the same year.

The Second World War brought about massive needs for military transport aircraft to quickly move both men and equipment, so the design was modified yet again with a strengthened floor, freight door and the engine power up-rated with the fitment of a pair of 1,200 horsepower Pratt and Whitney Twin Wasp radial engines. This new type was designated the C-47. Over 10,000 were built between 1939 and 1945 and, following the war, a great number were sold off cheaply by numerous Air Forces to the many fledgling airlines that emerged. These operators found the aircraft's rugged strength, large freight door and low operating economics perfect for their type of operation - and so it was with Dan-Air.

Mr F. E. F. Newman, founder and Chairman of Dan-Air recalls how they entered the airline business;

> "For the first few months, the aircraft continued to operate under the management of Meredith Air Transport, for not only did we not have the staff, we needed time to set up our own operating organisation. The aircraft was mainly flying between Berlin and Southend on what was known as 'the little airlift'. In the autumn the aircraft came under total Dan-Air control. Captain Garside joined us to run the flying side with Mr Selby Lownes running the commercial side. Mr Austin, the then Company Secretary of Davies and Newman dealt with financial and insurance matters and, together with the other Directors, formed a satisfactory team to commence operations in-house".

Mr F.E.F. Newman MC CBE, founder and Chairman of Dan-Air Services Ltd for 37 years of it's 40 year history.

As G-AMSU was already based at Southend, it was natural for its new operators to continue flying from there. The first commercial operation of Dan-Air's Dakota occurred in June, a charter from Southend, via Ringway Airport, Manchester to Shannon Airport with Captain Garside at the controls.

BIRTH OF AN AIRLINE

Take one aircraft... Dan-Air's first aircraft, the ex-Meredith Air Transport DC-3 Dakota G-AMSU at Innsbruck sometime in the 1950s (DAS/GMS collection).

Flying the 'Little Airlift'

For much of that summer and following autumn, the Dakota was gainfully employed on the second, or 'Mini' Berlin Airlift. This relatively little-known event had began in 1951and was a much more subtle attempt by the Soviet's to place an economic blockade of West Berlin's emerging industry. Their first attempt in 1948 and 1949 was targeted on basic living requirements; coal, food and the like. The second blockade began by impounding manufactured goods such as electrical components that were leaving the city along the rail and road links and earning vital revenue for the city. They also blocked raw materials from entering.

Another problem was the growing flood of refugees who were still able to cross into the western zones of the city, for the infamous 'Berlin Wall' was still ten years away. The Russians saw these dissidents as a means to help their aims, for they were an extra burden to the authorities. Dan-Air's single Dakota assisted in bringing out some of the many thousands of refugees that had fled across the border.

For the remainder of the year the aircraft operated numerous *ad hoc* charters to many parts of the British Isles and Europe. By the end of 1953 the single aircraft had carried 4,243 passengers and had flown 108,921 revenue ton-miles.

A second Dakota, G-AMSS, was obtained in February 1954 from William Dempster Limited, with other aircraft following close behind. This Company had been formed in 1949, making use of a pair of De Havilland DH89 Dragon Rapides obtained from Ciro's Aviation on passenger and freight charters. In 1950 a change of role came about with the purchase of a pair of Avro Tudors from British South American Airways, which allowed the Company to specialise in low-cost charter flights from London Blackbushe to Johannesburg in South Africa in close association with Pan African

DAN-AIR SERVICES

The second aircraft. Here DC-3 G-AMSS (possibly at Blackbushe) is still in the colours of William Dempster Limited, its former operators. That was soon to change...
(Friends of the DC-3)

Air Charter, a South African charter company. By 1951 the operating base had moved from Blackbushe to Stansted.

Throughout its life, William Dempster Ltd had been plagued with troubles - both financial and legal - so much so, that by the end of 1953 it had ceased flying. Dan-Air soon placed their surplus Dakota to good use.

Initially, Dan-Air's traffic was mainly cargo flights, but passenger charters gradually increased. The embryonic airline's aircraft movements for the time reflected the parent Company's involvement in shipping, for many Dan-Air charters were flown to quickly carry replacement ships' crews and urgent marine spare parts around Europe. It was through this aspect that the Company's aircraft became familiar sights around many European airports situated close to sea-ports. In late 1954 both aircraft found valuable additional work when a dock strike left goods stranded at sea-ports all around the UK; at the same time exporters abroad also found it impossible to get their goods into the country. Dan-Air an a number of other airlines flew cargo to many different destinations in Europe. The wartime spirit lingered on and the airline's small staff pitched in and did all they could to help.

The civil air transport industry of the time was very different than it is today. Aircraft and their associated equipment were technically much less complex and the rules and regulations governing them were much more simplistic, which admirably suited the temperament of many of the pilots and engineers who had learnt their trades in the heat of battle.

Financially, times were hard, but even in these early days there were two areas where Dan-Air never compromised; Flight Safety and Training. Those in the office may have struggled to find paper-clips, but everyone was fully trained to operate the equipment safely!

The fledgling air-package tour industry.
The early 1950s saw the emergence of an apparently new form of tourist travel, popularly named the 'package tour' industry (for usually a customer purchased a complete 'package' - travel,

BIRTH OF AN AIRLINE

... to the full 'Dan-Air London' livery. The large rear cargo door (with inset passenger entrance) is clearly visible in this photograph taken at Blackbushe. (DAS/GMS Collection).

accommodation and food - in one transaction). It is more correctly called an 'Inclusive Tour' (IT).

Although Thomas Cook had started his famous 'Cook's Tours' which were early versions of the IT, it is to a travel company called Horizon that credit must go for starting the first air travel-based Inclusive Tour. The year was 1949 and over that summer season Horizon arranged from its modest Fleet Street office for some 300 holiday-makers to fly out to the island of Corsica to make the first ever all-in holiday by air to a tented camp!

Cheap air travel had not been easily available to the general public before the Second World War, for although the basis for today's world-wide air transportation system was laid down by many of the airline pioneers of the 1930s, travel by air was then for either the very wealthy or for the businessman for whom time was money.

The period of the Second World War saw many people from the United Kingdom stationed overseas for long periods of time at His Majesty's expense. When the war was over and they were 'demobbed' they went home and told their friends and relations that a whole world existed on the other side of the Straits of Dover. This awoke a desire to see the world and, as they also had money in their pockets to spend, the travel agents were almost forced to respond to the demand! This new type of travelling public did not want to go to the trouble of making individual travel and accommodation arrangements, organising travel from one company and hotels with another - they just wanted to pay their money and go!

'Package Tours' as such had been available to the public for a number of years before the war, but they tended to be organised using coaches and railways, which placed a severe restriction on the distance that people could travel in a set period of time.

With the proliferation of surplus military transport aircraft lying virtually abandoned on airfields after the war, many small independent air companies sprang up, usually managed and operated by ex-Air Force aircrew who had been flying for a very different purpose only a few years before. Machines could be bought at knock-down prices and although work was available for the aircraft, it had to be fought for. Thus in this competitive environment, any travel agent that contacted a company for a price to fly a number of holiday-makers abroad knew that if one company failed to give him a good price there was always another on the other side of the airfield that would!

So gradually more and more members of the public became aware of the possibilities of travelling by air for their annual holidays, despite government applied restrictions on foreign travel and restrictions on the amount of currency that could be taken out of the country - set at £25 per annum in 1952, but raised to £40 in the 1953 budget! But passenger charters was not the only form of business available to the independents.

There was also *ad hoc* charter flying for the War Office, much of which related to long-term contracts. National Service, in which young men had to serve two years in the Armed Forces, was still in force and these young men were scattered all around the world on peace keeping duties. The idea of flying troops, equipment and urgent spares around the globe instead of sending them on long sea journeys by troopship had first been proposed by the Blackbushe-based Airwork Ltd in the late 1940s. Eventually the War Office became convinced and, in 1952, calculated that it could save £4 per head by flying troops out to the Middle East instead of sending them by sea Thereafter the air-trooping business exploded, generating much work (and capital) for many of the independents.

This form of flying had the effect of educating the next generation of working man that air travel was possible for all and, furthermore, it created more awareness in the general public's mind that travelling abroad was possible at affordable prices.

It was in this climate that Dan-Air found itself operating during its early days - the foundations for the concept of 'cheap air travel for all' were being laid down all around them, but no one involved at the time could have been aware of how the holiday travel market would explode in just a few years time.

CHAPTER TWO
The first move...

After eighteen months of successful operations from Southend, Dan-Air moved its base to Blackbushe near Basingstoke in Hampshire during January 1955. At that time the airfield at Blackbushe was a flourishing airport and was used by an increasing number of operators who wanted an easy and unencumbered entry and exit for London. Because of this, the airfield had evolved into one of the country's main aerial gateways, but unfortunately it suffered from what today would be a totally unacceptable operating problem - most of the airport, including the Passenger Terminal, was on the north side of the main A30 London to Salisbury road whilst the hangars and maintenance area were on the south side! Not only did motorists have to be warned of low flying aircraft but literally that aircraft would be crossing the road, for one of the runways and a number of taxi-ways actually crossed the highway!

Blackbushe airfield had been built in late 1941 as a bomber airfield for the RAF. The airfield was transferred to the Ministry of Civil Aviation and re-opened in February 1947 as Blackbushe Civil Airport. Much of the civilian effort in the Berlin airlift flew from here, although there was still a strong military presence - the United States Navy operated a large facility, and the RAF used the airport for many of its overseas trooping operations.

On the civil side, a number of independent airlines operated from Blackbushe - British Aviation Services, Airwork Ltd, Silver City Airways, Westminster Airways, Eagle Airways - names that are now long gone.

Possibly in order to avoid some of the problems associated with operations from Blackbushe, and certainly due to the lack of hangarage available there, Dan-Air started a search to find a home for Dan-Air Engineering Ltd, their embryo maintenance team. It seemed sensible to the management that in order to be successful in the commercial operation of aircraft, 'in house' maintenance for the growing fleet was essential. This was eventually located at the old wartime RAF airfield at Lasham in Hampshire.

The airfield at Lasham had a distinguished and meritorious history. Opened in the autumn of 1942, the airfield remained unused until March 1943, when 175 Squadron of the RAF moved in for a few days. It was from this airfield in February 1944 that De Havilland Mosquitoes of the 2nd Tactical Air Force set out to bomb Amiens Prison in order to free French Resistance workers.

The airfield was retained in the post war years, becoming an inactive site in Fighter Command. In the early 1950s permission for

DAN-AIR SERVICES

gliding was obtained and the airfield quickly became very popular. The National Championships were held there in 1955, just at the time when Dan-Air was looking for a maintenance facility.

Mr Newman remembers the search for a maintenance base with a high degree of clarity;

> "We had just bought three Yorks and had nowhere to put them. As luck would have it we came across a disused airfield nearby at Lasham - then a satellite to Odiham. Overall, the airfield was in excellent condition, for the runway had just been resurfaced following the Berlin Blockade and the heightening of the Cold War. On the field was a large empty hanger, that had been used the previous winter by a circus and came complete with elephant droppings on the floor!
>
> The Air Ministry granted us permission to use it under licence, but it was not until a number of years later, after long and tedious negotiations that a 21 year lease on satisfactory terms was granted."

Avro's mighty York!

Dan-Air's move to Blackbushe coincided with the withdrawal from service of the RAF Transport Command's fleet of Avro Yorks, resulting in the first of three ex-RAF aircraft joining the Dan-Air fleet in early 1955 - one of six such machines that were eventually used, the others coming from civilian sources.

The Avro Type 685 York had been designed as a transport derivative of another of the Company's products - the famous Avro Lancaster bomber - and was largely a private venture produced under the difficulties imposed by an wartime Anglo-American agreement that compelled British firms only to build combat machines.

The York made use of a number of Lancaster assemblies - mainplanes, powerplants, undercarriage and tail units, although a completely new 'box' fuselage was designed and built that, because the main spar of the wing now passed through the roof, sat much lower to the ground for ease of loading passengers and freight. By locating the main spar at high level the designer, Roy Chadwick, ensured that the main cabin remained free of obstacles. Although the prototype made its first flight as early as July 1942, few materials were then available for production, so full quantity production did not start until 1945, after which over two hundred were built for the RAF and civil airlines.

It was from Blackbushe that Dan-Air's Yorks operated the long-distance charter flights, mainly carrying cargo to Africa and the Far East, but also flying a series of trooping flights to the Middle East. John Kew, then serving as a cook with 6 Sqn RAF, has vivid memories of one such flight out from Blackbushe on 21 December 1955.

> "I remember the date clearly, for it was my wife's birthday! We left Blackbushe and night-stopped at Malta for fuel. The noise in the main cabin from the four Rolls-Royce Merlins made any form of in-flight announcements impossible, so we made do with the occasional note of our position passed around on a slip of paper.
>
> The next day we flew on to Nicosia, Cyprus and then on to Habbaniya in Iraq where I spent a number of months before moving back to Cyprus.

THE FIRST MOVE

Avro York G-ANTK (wearing the later colour scheme of the outer tails in dark blue with red tips and stripe for the registration), comes into land at Ringway Airport on 9 September 1961. The tails were painted dark blue to disguise the oil and exhaust stains that were the inevitable result of being placed directly downwind of the Rolls Royce Merlins.
(Paul Tomlin)

At that time it was an experience for anyone to fly. Looking at the couple of survivors, the York may seem small now, but at the time it seemed enormous. And the stewardesses - remembering back they seemed gorgeous to a young airman away from home for the first time!"

Another passenger who remembers well those trooping flights of the 1950s is John Burch:

"Whilst I was in the RAF (1949-73) I was posted out to the Canal Zone in Egypt. My draft assembled at Lytham St. Anne amid much speculation about which ship would troop us out. To our surprise, we were bussed to RAF Lyneham and went out by Skyways York!

Whilst in Egypt I worked for a while on the Transit Aircraft Flight (TAF) and we regularly used to get Dan-Air aircraft passing through. One incident of note that comes to mind involved one of their Yorks coming in to land at RAF Fayid. Near the airfield was a small hill, about 200 to 300 feet high, called Mount Shubra. This obstruction was normally highlighted at night by paraffin-fired goose-neck flares, but on this occasion, the flares had been stolen, and so Mount Shubra was lost in the darkness to the approaching aircraft which just clipped the top of the hill, landing with the starboard outer propeller blade tips bent right back. None of the passengers realised what had happened!

When the Dan-Air engineer arrived to change the prop and shock-test the engine, he used whatever RAF tools were available locally - and anyone willing to lend a hand. On completion of the work, all his 'helpers' got a £5 back-hander and a dozen bottles of Stella beer - the going rate for the job!

Another perk on the TAF was helping to clean out returning aircraft. We used to look forward to the Dan-Air (and other airlines) aircraft, for once the passengers had disembarked, we could get our hands on that day's newspapers and whatever fresh English milk there was left in the galleys!"

DAN-AIR SERVICES

Looking surprisingly light and airy is this interior of a Dan-Air York. Note how the floor (and therefore the seat lines) had a distinct curve and the overhead 'obstruction' half-way down the cabin - this was where the main spar of the wing passed through. The aisle is also so narrow that a galley trolley of today would never pass down it! (DAS/GMS Collection)

It was not only servicemen that went on these trooping flights but their dependants too. Mrs Joan Peake flew out to Malta in 1956 to join her husband who was serving in the Royal Navy there:

> "This was my first flight, and I was travelling with two babies, aged 18 months and six months. The flight was full of Navy personnel, but I was amazed by the kindness shown by the cabin crew (but they were called air hostesses then). I remember we stopped at Nice Airport to re-fuel and we actually left the plane for a couple of hours. One thing that has not changed is the in-flight meals on their individual trays - at the time I found this quite extraordinary! Incidentally, since then I have flown many thousands of miles, but have just returned from Venice - only my second flight with Dan-Air!"

The Dakotas were now undertaking a whole series of passenger charter flights, which included a high number of international IT flights flown out of the airport. Many passengers were surprised to

A pair of Dan-Air Stewardesses pose either side of Nevil Shute Norway, the world famous author of 'A Town Like Alice' and 'Slide Rule' that detailed his life as an aviation designer, about to board a Dan-Air DC-3. On the left is Deirdre McRae and the girl on the right is thought to be a Ground Stewardess because of the lack of 'Wings' on her jacket. The photo must have been taken before 1960, when Nevil Shute passed away in Australia. (DAS/GMS Collection)

THE FIRST MOVE

see the flight crew walking out to their aircraft on even the sunniest of days carrying raincoats. This was not because the 'Met' had forecast rain at their destination; more that the DC-3 cockpit leaked like a sieve and it became a common practice for pilots, when flying through cloud, to drape the raincoat over their knees to keep the drips off their uniforms!

The flights to Calvi continued every Sunday from May to September on behalf of Fregata Travel. Flights were also operated on behalf of other travel agencies to destinations that included Biarritz, Ostend, Paris, Perpignan, Pisa and Tarbes.

In June 1956 one of the Dakotas opened the airline's first scheduled passenger service - a summer only service that linked Blackbushe with the island of Jersey.

War Office contracts

During 1956 Dan-Air obtained an Air Ministry contract for the carriage of stores from Lyneham to Singapore. To operate these flights, a further two Yorks were obtained from Freddie Laker's Stansted-based company, Air Charter. This contract, along with others, kept Dan-Air's Yorks gainfully employed for a number of years, plying regularly throughout Europe and the Middle East to Asia and Singapore.

During one flight back from Australia, John Cameron, Dan-Air's Chief Pilot (who had taken over from Captain Garside) met up in Singapore with Captain Arthur Larkman, who at that time was flying with Malaysian Airways. Arthur, was to join Dan-Air at John Cameron's request in September 1956, later became the Director of Operations in 1977 and eventually retired in early 1988. He remembers that meeting well...

John Cameron, the second Chief Pilot of Dan-Air, seen in front of a Company York with the Wallaby presented to him during an enforced stop-over due to engine problems at Alice Springs, Australia. Just visible is the Wallaby emblem on the nose of the York. (DAS/GMS Collection)

"I was sitting in a hotel bar ordering drinks whilst John was ordering dish after dish of salad, much to everyone's puzzlement. It seems that during an enforced stop at Alice Springs on the way back, due to engine problems with his York, he had been presented with a wallaby... and as he was keeping the animal in his hotel room, he needed the salad to feed it! Back in England much fuss was made of the airline's new mascot and its arrival made all the papers. The Wallaby was presented to a zoo in Sussex where I believe they now have a colony".

It was from Lyneham that Dan-Air picked up another member of staff who was to play a very important part in the development of the airline. Alan Snudden was a Loadmaster on the airfield in charge of the aircraft loaders. Captain Larkman can clearly recall what happened...

"Civil aviation was (and still is) a very small world - you soon got to know the strengths and weaknesses of those in it. In fact, that was very much an on-going part of Dan-Air's success, for you often had personal knowledge of those you were taking on. We had been impressed with Alan's efficiency in sorting out whatever problem arose, so as he was due to leave the RAF anyway, we persuaded him to join us as a traffic officer at Blackbushe. From there he moved up to Head Office in London..."

DAN-AIR SERVICES

Alan Snudden eventually became Managing Director of Dan-Air Services and then moved to Luton with Monarch Airlines.

On 16 November 1956, following the invasion of Hungary by the Russians, the Hungarian refugee airlift started and Dan-Air was quick to become involved, providing its first flight free of charge. Over the next four weeks two of Dan-Air's Dakotas made eleven flights from Blackbushe to Lintz and Vienna to collect 31 or more Hungarians every trip. By the time the airlift ended on 14 December, 350 people had been flown out.

Specialised freight aircraft.
In 1957 two new aircraft types were introduced into the airline's service - a Bristol Freighter (G-AINL) was acquired from the manufacturers for use on the long-range freighting services and Government freight contracts, and in June a De Havilland Heron was leased from Overseas Air Transport for six months to be used on the passenger services.

From Traffic Officer to Managing Director! Allan Snudden. (A. Snudden)

The Bristol type 170 Freighter was a transport aircraft originally designed to be capable of operating from jungle airstrips and was, therefore a rugged, high-wing monoplane, built solely for low-initial cost and cheapness and simplicity of operation. Expensive alloys were dispensed with in favour of steel and the 2,360 cubic foot square section fuselage was free from all obstructions. The flight deck sat on top, reached by an internal step-ladder. The wing was fitted with hydraulically-operated split flaps and carried interchangeable 1,650 horsepower 'Hercules' piston engines in

A dark and cold 16 December 1956 - some of the 350 Hungarian Refugees rescued de-plane from a Dan-Air DC-3 at Blackbushe. Most escaped with very few possessions and the clothes they were wearing. (DAS/GMS Collection)

THE FIRST MOVE

A Bristol Freighter is loaded with cargo. The open nose doors show how bulky cargo could be loaded direct from ground transport and the sturdy, fixed undercarriage is clearly visible. (DAS/GMS Collection).

underslung nacelles, from which was also attached the fixed undercarriage. A second Bristol Freighter, G-APLH, arrived at Blackbushe on 31 March and operated its first service, from Blackbushe to Gibraltar.

1959 saw a change for the airline when the Air Ministry freight contract to Singapore came to an end, but the Company was awarded a replacement - the carriage of the experimental *'Black Knight'* rocket with its associated personnel and equipment from the United Kingdom to the rocket range at Woomera in Australia for testing. The Freighter was the only British aircraft capable of carrying the rocket and even then, Dan-Air pilots had to take a rather roundabout route. Due to the secret nature of the rocket, a number of countries that insisted on all cargo being fully inspected had to be avoided at all costs. Arthur Larkman made eleven such flights, each taking on around 170 hours flying time.

"The rocket, packed in a enormous box, appeared on the manifest as 'Agricultural Machinery' and absolutely filled the aircraft. We had to avoid Egypt of course, but the biggest problem was Indonesia, which we could not cross. So we headed out into the Pacific to the Phillipines, down through New Guinea and then south-west to Darwin. That was a long way in a Bristol Freighter, and so we had an extra 100 gallon tank fitted into the nose doors, from which we pumped fuel up into the wing tanks!

For the first few flights we went through Indo-China and on one such flight I had to turn back to Saigon due to an engine problem. At that time there was an International Control Commission - Polish, Russian, all kinds of nationalities - on the airfield, and there we were. I rang up the British Consul and explained that we had a problem and could he help in arranging the loan of an engine from a New Zealand fighter squadron based in Singapore? He replied 'We do not deal with charter airlines old boy'. So I got into a taxi and told him face-to-face what we had on board; we soon got everything we wanted to get us on our way in double-quick time!

DAN-AIR SERVICES

Two views of the Malta incident. Above: 'MUT with it's nose 'in the weeds' at the end of Luqa's runway. Note how close the flight deck is to the runway approach lights and the wrinkled fuselage just forward of the airlines emblem.

Below: A rear view of the same incident. The earlier colour scheme of all-white tail with a red band lined in blue is clearly visible, as is some damage to the rear fuselage.
(Peter Carter Collection)

1958 was the year when Dan-Air suffered its first accident. The Company had been in existence for almost exactly five years without a major incident when over the space of five days it lost two aircraft, both Yorks. The first was when G-AMUT slid off the end of the runway at Luqa, Malta on 20 May, luckily without any serious injury to those on board. Five days later 'UT's sister aircraft G-AMUV suffered a serious engine fire and force-landed at Gurgaon, India. Only the radio operator, Jack Maloney, escaped. Both aircraft were written off and, after another accident to one of Air Charter's Yorks, from then on the type was only allowed to be used as freighters.

The Middle East was an area that could always be relied upon to throw up one incident or another. Pilots (and Navigators) of the freighters had to take particular care when flying over the mountains of north-eastern Turkey, for the aircraft had to pass through a narrow corridor of airspace bordered to the north by Russia and to the south by Iraq. If the radio signal from the Lake Van beacon was not correctly interpreted, they ran the risk of being shot down if they strayed from the corridor.

There were other colourful incidents also. Dan-Air were the first independent company to send an aircraft back into Alexandria after

THE FIRST MOVE

The story behind this picture is interesting, for it is a fake! All the occupants are not happy passengers, they are Dan-Air staff - the Stewardess is Maurene Thompson and the couple in the second row are Maurene Breckall (a secretary) and Brian 'Sport' Martin, one of Dan-Air's pilots! The photo was taken as a publicity shot to show the interior of one of the Bristol Freighters that had been converted to carry passengers. (Maurene Thompson).

the Suez crisis; the pilot claimed afterwards that the only reason the Egyptian Air Traffic Control let him land was because they thought the aircraft belonged to a Danish company! By the time they found out, the pilot had received a warm personal welcome as a New Zealander, so all was well!

Scheduled Freight Services

The British European Airways Corporation (BEA) awarded Dan-Air a two year contract to fly its London-Manchester-Renfrew (Glasgow) scheduled freight service. On the evening of 25 May 1959 Dan-Air's DC-3 G-AMSS left Heathrow for Renfrew and then returned to London via Manchester. For the remainder of the year this service was flown nightly by the Dakotas, but by the end of the year the Yorks began to appear on the service as they became available after release from the Air Ministry contract work.

Due to the low annual utilization of the Bristol Freighter fleet - at about a thousand hours per aircraft - at least one of the Freighters was converted into the 44 seat '*Wayfarer*' passenger configuration for the 1959 summer season. Along with the Dakotas, this aircraft was placed on the airline's many passenger services radiating from Blackbushe to the continent.

Apart from these services, the passenger aircraft were also used on the Company's first Inclusive Tour charter flights from Manchester to Basle, Ostend, Lyons, Palma, Pisa and Zurich.

The airline also undertook a small Inclusive Tour programme from Gatwick on behalf of Motours. Every Saturday morning the

DAN-AIR SERVICES

aircraft would fly a party of tourists to Munich or Nice and then return to Gatwick early in the evening with another party, before positioning the aircraft back to Blackbushe. Then on the next two Saturdays the aircraft would fly to Nice, and the following two Saturdays to Munich. In the meantime the tourists would find a car waiting for them at the airport which the could then drive at a leisurely pace to the other airport for their flight home. On these flights the aircraft could carry as many as 32 passengers and by the end of that summer season aircraft utilization was around 1,300 hours per machine.

1960 was a similar year of change. By the spring the Yorks had almost completely taken over the London (Heathrow) - Manchester (Ringway) - Glasgow (Renfrew) freight service started the previous year. Because of their success, Dan-Air was awarded several other contracts from British European Airways, including a weekly London to Milan and London to Rome service and, during June, a weekly London to Brussels service. The Ministry of Supply support services to the Far East and Australia continued, but even with this high level of usage, the Yorks still found time to fit in the odd *ad hoc* charter.

During the 1950 and 60s the UK Government had a number of export drives. As can be seen from this advert, Dan-Air did its bit!

Keeping 'em flying...
The times and working conditions for the ground engineers during the mid-1950s were hard, Stephanie Dowling (now Coley) was one of the first females working for the airline in a technical position, employed as a radio mechanic in the Lasham radio workshop under Chief Radio Engineer Ray Climas:

> "I joined the Company from the Wrens in 1958. There were just four of us in our workshop, all doing repairs and maintenance on the fleet's radio equipment. Although it was only a small department, it was a very happy atmosphere, although at times it could get very hectic when a job had to be done in a hurry!
>
> As one of the first females working for Dan-Air in a job that was not administrative or secretarial, I have always regarded myself as one of the trail-blazers for the women pilots that came along after I left towards the end of 1961!"

Peter Carter (latterly Commercial Manager, Aircraft) remembers. Peter had left school and started work immediately with Airwork Limited at Blackbushe, but then moved to Dan-Air:

> "The hangars we used were draughty and, apart from glowing coke braziers, they were unheated in the winter. If a fault developed with any of our aircraft at Blackbushe, or whilst it was away from base, we had to move 'heaven and earth' - even resorting to the tried and tested principal of 'beg, borrow and steal' to get it back into service. There was a saying common amongst the Dan-Air ground engineers in those days after they had made a repair; *'Don't worry, it'll do for a trip'!*
>
> Nowadays when an aircraft goes 'Tech' we just ship parts around the country on other airlines or, if there's not an aircraft going that way soon, hire a taxi to get the part there quickly. It was all a lot different then.
>
> I remember one Good Friday when reports came down to us that we had a DC-3 stuck up at Newcastle with an undercarriage unit that would not retract. Compared

THE FIRST MOVE

to to-day's aircraft, the DC-3 was (and still is) a very rugged machine and simple to work on. The undercarriage mechanism relied on a single hydraulic retraction jack on each leg to move it.

So on this particular Good Friday what I had to do was draw a spare jack assembly from the stores and, still in my oily overalls, get a lift down to Heathrow's Terminal One, hoist it onto my shoulder and, with my tool-kit in my other hand, queue for check-in on the next flight north!.

When I got to the Check-In Desk, a beautifully turned-out BEA ground hostie girl on duty asked me what I was carrying. I explained and told her to weigh it. She tried to take it from me, but it was so heavy that she literally dropped it on the scales! Of course, the shock of this made the ram shoot down and she got covered from head to toe in hydraulic oil! I apologised, lifted the jack back on my shoulder and carried it through the terminal and onto the aircraft!"

Operations from Bristol

By 1959 the Company felt that it had accumulated sufficient experience to operate scheduled passenger services on a much larger scale. This coincided with Dan-Air being approached by the local authorities who operated the newly-opened regional airport at Lulsgate, located adjacent to the A38 trunk road some seven miles south of the city of Bristol. The airport was formerly RAF Lulsgate Bottom, opened as a training base during August 1940. As with many RAF airfields, it was transferred to the Ministry of Civil Aviation after the war, but development into Bristol's future airport was painfully slow. Finally the site was purchased by the Bristol Corporation in 1955 and the Airport was opened by Princess Marina in May 1957 to replace the pre-war Whitchurch airfield.

A number of services using De Havilland Herons were established at the new airport by Cambrian Airways, the Cardiff-based Welsh airline, but in the winter of 1958, a recession in the air travel business forced Cambrian to cease flying for a while and put their entire fleet all but one aircraft up for sale. This left the expensive new airport at Bristol with only one regular weekday flight - an Aer Lingus service to Dublin.

The council asked Dan-Air to set up a network of services to link the airport with other cities, both within the United Kingdom and in

A pair of Dan-Air DC-3s (G-AMPP & 'MSU) at Bristol in the company of an Aer Lingus F.27. (DAS/GMS Collection).

DAN-AIR SERVICES

DH Dove G-AIWF that was used to start the Bristol to Liverppol service via Cardiff. (Mike Forsyth Collection)

Europe. As a result, Dan-Air were approached by the ATAC to establish scheduled services, linking Bristol with Cardiff, Liverpool and Newcastle, for which approval was granted.

Interestingly, the initial application for these routes mentioned the American-built Convair 340 as one of the types to be used, but in the event, in order to operate these routes out of Bristol, Dan-Air acquired a pair of De Havilland Doves in January 1960.

The De Havilland DH104 Dove was an all metal, low-wing, twin engined eight seat passenger aircraft designed as a modern replacement for the Company's pre-war DH89 Dragon Rapide biplane airliner design. Power for the new aircraft was supplied by a pair of De Havilland Gipsy Queen 71 supercharged, inverted six cylinder piston engines, each fitted with fully feathering and reversible pitch propellers. Dan-Air Services' examples were both obtained second-hand, G-AIWF coming from the College of Aeronautics at Cranfield and G-ALVF from Hants and Sussex Aviation Ltd.

On 4 April, Captain Lancaster lifted Dove G-AIWF from Lulsgate's tarmac to inaugurate the year-round service to Liverpool and Cardiff. Initially three services each week were flown and on 14 May Captain Brian 'Sport' Martin flew G-ALVF for the first trip on the scheduled summer-only route to the Isle of Man. 'Sport' Martin was later to become a legend within Dan-Air. He got his nickname from his hearty greeting of *"Hello Sport!"*

De Havilland Heron G-ANCI at Lasham at the start of its lease from Overseas Air Transport. (GMS/DAS Collection)

THE FIRST MOVE

Another view of Heron G-ANCI, this time clearly showing the fixed undercarriage and four Gipsy Queen engines. (Mike Forsyth Collection)

Introduction of these Lulsgate-based services led to Dan-Air acquiring hangar space on the north-west side of the airport and setting up a small maintenance base to look after the Doves there but with facilities large enough to take Dakota-sized aircraft should the services eventually require their use.

The first international scheduled service to be opened by the airline began on 16 July 1960, when Dakota G-AMSU flown by Captain Davies opened the seasonal Cardiff - Bristol - Basle passenger service, flown throughout the summer by the Dakota fleet. Also in July the airline extended its Liverpool - Bristol service to include Plymouth. Thus the airline's West Country network became well established, and this was to provide a strong foundation for its famous 'Link City' network.

Expansion and Appointments

By now things had become so busy it was necessary to have a General Manager, so in 1955 Mr Laurence Moore moved across to fill the new position from Cambrian Airways.

Laurence ('Dinty') Moore had been Cambrian's Chief Engineer. He later became Managing Director of Dan-Air until he left the Company in 1963.

Mr Moore wasted no time in recruiting Bryn Williams from Freddie Laker's Air Charter, as Dan-Air's Chief Engineer and he reigned over the Engineering Division until he retired in 1980. Shortly after this, Mr Frank Horridge, formerly of William Dempster, joined as Chief Inspector and thus formed the team which successfully masterminded the engineering and technical side of Dan-Air over the years.

Bryn Williams (right) and Frank Horridge (far right), for many years masterminds of the Dan-Air's engineering and technical side. (DAS/GMS Collection)

DAN AIR SERVICES

As well as a successful engineering division, it was essential to have a safe and reliable operations team. Amongst the many pilots, the airline was fortunate to recruit Captain R. A. Atkins (known to his colleagues as Bob) who joined Dan-Air from the RAF in November 1955. Captain Atkins was to become the airlines Chief Pilot in 1963.

Back at Head Office, it was necessary to enlarge the Commercial Department, and Mr R. A. Pigeon took over from Mr Selby Lownes, who had left the Company.

One of the very early cabin staff members was Scottish born Maurene Thompson (known to all as 'Haggis'), who went along to Blackbushe Airport one day in 1958 to fulfil her childhood dream - she wanted to become an Air Hostess. She had already prepared herself, learning some Italian (one foreign language was a required skill with the larger airlines) and had already had a years training as a nurse:

Dan-Air Stewardess Maurene Thompson, who later married David Charlwood, a Dan-Air First Officer. While working for Dan-Air she was universally known as 'Haggis' and is seen here in the doorway of Bristol Freighter G-ALPH.
(Maurene Thompson)

THE FIRST MOVE

"I went for an interview with Eagle Aviation (the forerunner to British Eagle) and while I was waiting I struck up a conversation with a rather 'posh' but smart little girl. *"When we've had our interview here..."* said the English girl *"...why don't we go over and try Dan-Air? I hear they need girls too"*.

We found Dan-Air in a wooden hut which served as a Chief Pilot's Office, Crewroom and local HQ. Around the place were a number of lean, bronzed Australian-sounding men in World War Two type flying boots, jerkins and peaked caps - the absolute epitome of flying and just the sort of pilots I imagined would be flying the airways!

We were both interviewed by an Australian, Warren Wilson, who suggested we then went to Head Office at Bilbao House near Liverpool Street Station in London. Here we were seen by Alan Snudden, the Managing Director, and I was taken on at £8 per week on a temporary basis for the summer season.

Back at Blackbushe we were given practical training and Val Fraser, the Chief Stewardess, who lectured us sternly on how to avoid co-habiting with the pilots whilst away on night-stops and gave us strict instructions as to how we should behave. She put the fear of God into us - so much so, that for the first few trips away I used to barricade my door, but I soon realised that it was all totally unnecessary.

One thing I remember particularly well is the difference between the aircraft galley's then, compared with those fitted to todays modern aircraft. There was just a small sink with cupboards above so as to be able to store dry stores and a worktop with cupboards underneath. The hot water came from a heater fixed onto the front galley bulkhead. This could be topped up from another container above. The guages for the water level was always playing up and you could never ascertain how much water you had left! Ice for the drinks was placed aboard in a large ice flask. The bar boxes were free standing and we had to be careful how to stow them out of harms way for take-off and landing. Bar and meal trolleys were non-existent - everything was brought to the passenger by hand.

There was no facilities on board for heating food, so it was generally egg or chicken salads, tinned fruit and two water-biscuits. On some aircraft we could use hot cups; these were flasks that could be plugged into the aircraft's electrical system and were very popular. We could rustle up scrambled eggs, soups... very handy on a cold day! Coffee and tea was served from blue plastic crockery, accompanied by a great big smile from the stewardess!

After my first summer I went back to London for another interview with Alan Snudden and, to my surprise I was offered a permanent job - the only new arrival to be kept on!"

The first recorded use of cabin staff by an airline in the United Kingdom occurred in the early 1920s when Daimler Airways appointed a number of 14 year old boys to act as 'Flight Attendants' They were specially trained by the Savoy Hotel in London and wore full Page Boy uniforms. However, their use was restricted, mainly because they were forced to remain seated during the flight!

In March 1930 Ellen Church, a qualified nurse and licensed private pilot, wrote to Boeing Air Transport (BAT) in the USA and changed forever the passengers' view in the main cabin. Ellen suggested that BAT would benefit if they employed suitably qualified ladies (just like herself) to act as cabin attendants. The airline's bosses were shocked at the thought of women flying as crew members, but were quick to appreciate the commercial value of the idea. Applicants were carefully vetted, with particular attention paid to nursing experience, age and stature of less than 5'4", the latter being so that they avoided banging their heads on the low roof!

DAN-AIR SERVICES

On 15 May 1930 Ellen finally greeted her first eleven passengers on board a Boeing 80A for a flight from Oakland, California to Cheyenne, Wyoming - a five sector flight that took nearly 24 hours. As part of her duties she had already cleaned the cabin, helped push the aircraft out of the hangar and had filled the fuel tanks! In some respects, things may not have changed all that much...

Maurene remembers that there were some tremendous characters flying with 'Desparate Dan' as they were called by some at the time...

> "Warren Wilson, who I met when I visited Blackbushe looking for a job, I soon discovered was a typical Aussie with excellent flying ability. Every now and again he would loose his temper and sack everyone, but was sweetness and light the next day!
>
> I used to fly regularly with Warren and Pete Skinner (or was it Skingley?). They were both killed soon after they left Dan-Air to fly for a Middle Eastern airline; a bomb was planted aboard their Dakota which exploded in mid-air...
>
> Then there was Mike Cowper! He was incredibly well-prepared for any journeys to remote places - he wore three wrist watches telling him BST, GMT and Destination time and he often used to carry three different brief-cases with him; one contained his papers and normal food to eat in spare moments, one which he classified as his 'Emergency Food Pack' in case of a landing in the desert at some remote spot and finally his 'Ditching Case' in case of a water landing!
>
> One of the Ground Engineers was Jock Christie who would often fly with us and look after the aircraft on turn-arounds. He would teach us about the aircraft and warn us not to throw lighted cigarettes down the 'Elsan' (the forerunners to today's highly sophisticated toilets) - if we did, the loo was liable to explode!
>
> I was also taught how to secure the aircraft after landing by putting in the tailplane locks etc. I also used to make up the Nav packs in the office prior to the flight, checking that all the AERAD charts were put in for the route. Often at Blackbushe I was asked to sit in the cockpit and use the brakes whilst it was being towed away to be parked over on the far side of the road".

In 1959 the Minister of Aviation decided that Blackbushe was of no further use to the overall London airport scene and would therefore close. Following this announcement, the airline started looking for a new main operating base - this time moving to an airfield that was to become its main home for many, many years...

CHAPTER THREE
A future at Gatwick!

During the early 1950s the nation's capital was served by four airports - London Heathrow (at the time still incomplete, even in its most basic form); Croydon (by now firmly established in its declining years and becoming surrounded by industry and housing that made any expansion impossible); Blackbushe (regarded by some as too remote and by many as unsuitable due to the road passing through the site); and the grass airfield at Gatwick.

There had been an airport in the Gatwick area since August 1930, when Ronald Waters and John Mockford obtained the airfield's first operating licence for a site adjacent to the race course to be operated by Home Counties Aircraft Services. Much development took place throughout the war, although there was an uneasy alliance between military and civilian users who shared the field, but with peace the grass strips reverted to commercial use in September 1946.

With the rapid expansion of travel by air, the Ministry of Civil Aviation decided that some of the capital's airports needed modernising to cope with both the increase of use and the new machines that were then about to be placed into service by the airlines. There was also a need for a firm policy of rationalisation of airport sites. After much consideration, a decision was announced in 1952 that the Gatwick site had been chosen for London's 'new' second airport and, in March 1956, the airfield was closed to allow total reconstruction to commence.

Two years later, and with much fanfare (on a massive site slightly to the north of the old airport and directly connected to the nation's railway network) the re-constructed airport was opened by Her Majesty Queen Elizabeth II on 9 June 1958 as London's second airport. There was a single 7,000ft concrete runway, a new aircraft parking apron and new terminal buildings with the then revolutionary 'finger' walk-way so that passengers could proceed under cover all the way to their aircraft.

Continuing its rationalisation policy, the Ministry of Civil Aviation announced in 1959 that Blackbushe Airport would close from midnight on 31 May 1960. Consequently Dan-Air prepared for yet another move. Space at Heathrow was at a premium, so it had to be the relatively under-used Gatwick. The airline had already flown a considerable number of services from there, so during May more and more business was transferred to Gatwick. The move was finally complete on the last day of airport operations, when the last aircraft, Bristol Freighter G-ALPH, moved out of Blackbushe.

DAN-AIR SERVICES

During the early years of operations from Gatwick, Dan-Air (along with British Caledonian) were to be the saviours of the airport for, in many respects, it was these two airlines that kept the airport going with their many charter and scheduled flights.

Stewardess Maurene Thompson remembers well her first flight into Gatwick...

> "We landed in a gale that was blowing well across the single runway! The Captain of our Dakota had to make three approaches to get down safely. When down, we taxied up and parked in front of the brand-new, glass fronted Terminal Lounge. Just a that moment there was a tremendous crash audible even from inside the aircraft - the wind had blown in the huge glass window. Dan-Air and 'Haggis' had 'arrived'!
>
> Despite the move to Gatwick and the airline's evolvement into a much more professional set-up, there were still vestiges of the old pre-move Dan-Air. I still kept a toothbrush and change of undies packed in an overnight bag that went everywhere with me - we never knew where we were going to end up!

Airspeed's 'Ambassador'.

November 1959 marked the arrival of the airline's first pressurised aircraft. With an eye to further expansion, Dan-Air purchased a fleet of three Airspeed Ambassadors from the Australian airline Butler Air Transport via aircraft brokers Shackleton Aviation Ltd. These aircraft had been used in the harsh environment of the Australian outback, a task for which they were not suited. Thus the machines returned to European skies, and entered service with Dan-Air on 15 March 1960, still in their original 49 seat passenger layout as adopted by Butlers when they obtained the aircraft from the British European Airways Corporation.

The Airspeed AS57 Ambassador had been designed as a Dakota replacement that could stage between European capitals with ease, carrying twice the load of the Dakota. It was a large metal high-wing monoplane, powered by a pair of 2,625 horsepower Bristol 'Centaurus' piston engines. Orders were placed by several airlines,

With both Centauri thundering at full power, Ambassador G-AMAE climbs out from Lasham after undergoing maintenance..
(DAS/GMS Collection).

A FUTURE AT GATWICK

including British European Airways, which placed the type in service under the name of the 'Elizabethan' class. Despite the undoubted passenger appeal of the Ambassador, BEA only kept the type in service for about six years, the design being replaced with the new turbo-prop Vickers Viscounts in 1957.

Introduction of the Ambassador allowed Dan-Air to further expand its Inclusive Tour network, so that during the 1960 season the Ambassadors were used on holiday services from Gatwick to Amsterdam, Brussels, Le Touquet, Munich, Nice, Santander, Tarbes and Tours. In addition they were used on many *ad hoc* charters and trooping flights to Berlin, Frankfurt, Gibraltar, Hamburg, Stockholm and Malmö from Gatwick. The airline was on course to become the largest operator of the Ambassador after British European Airways.

Into the new decade!

The move from Blackbushe to Gatwick saw the airline's one scheduled service from Blackbushe to Jersey move also, and this was re-opened on 18 June when Ambassador G-ALZX started the Gatwick to Jersey service. This service was flown throughout the summer by a combination of Ambassadors, Dakotas, Doves and Bristol Freighters.

Whilst the Ambassadors were settling down well in their new task of flying holiday-makers around Europe, the remainder of Dan-Air's passenger fleet were also well occupied in tourist travel. The Dakotas (G-AMSS and G-AMSU) spent a particularly busy summer transporting tourists from Gatwick to Barcelona, Basle, Biarritz, Calais, Deauville, Gibraltar, Groningen, Nice, Ostend, Stuttgart, Toulouse, Tours and Zürich. The aircraft were also used on the Manchester to Basle and Ostend routes, but mostly these services from Manchester were flown several times a week by the Bristol Freighters G-AINL and G-APLH, which were also used on a considerable number of summer charters.

The Bristol Freighter fleet was still under utilised however, despite the charter work and the BEA freight contract from London

One of Dan-Air's first Ambassadors, G-ALZX, seen here at Blackbushe before the move to Gatwick.
(DAS/GMS Collection)

DAN-AIR SERVICES

Heathrow to Milan, Rome and Brussels, so the airline applied to ATAC for permission to open a vehicle ferry service to the Isle of Man from Blackpool, Liverpool and Valley. However, this application was withdrawn later in the year when additional sea-ferries were placed on the routes, so no vehicle ferry service was ever flown by the airline.

1960 was thus a very important year for the Company, for it saw the commencement of a substantial network of scheduled passenger services, the introduction of the Ambassadors, the move to a new base at Gatwick and a significant increase in the amount of Inclusive Tour work flown.

1961 saw further expansion when, on 4 January, Dove G-AIWF extended the scheduled Plymouth - Cardiff - Bristol - Liverpool service into Newcastle, a route that was to be flown three times a week. Also in January of that year, Dan-Air had bought the aircraft and routes of Scottish Airlines (Prestwick) Ltd from Scottish Aviation. With this take-over the airline acquired a passenger-equipped Dakota and, most importantly, a licence to operate a scheduled service from Prestwick to the Isle of Man. This service was started on 27 May, the first of many flights that the company operated from and to Scotland. The Isle of Man was beginning to enjoy a very good air service from the Company, for in 1961 the island was also served with flights from Bristol, Cardiff, Plymouth, Exeter, Staverton and Swansea!

Dan-Air had operated a number of charters from Prestwick, and 1961 was no exception, with a number of mainly trooping charters from, of all places, Dusseldorf to Benbecula, stopping at Prestwick for customs clearance.

Passenger loads carried on the Plymouth - Newcastle stopping service reached such high levels that on 3 July Dakotas were introduced to cope with the demand. On 7 July, G-AMSS inaugurated a Bristol - Liverpool - Newcastle - Dundee route; Dan-Air used *HMS Condor*, otherwise known as Royal Naval Air Station Arbroath, as its Dundee terminal. Later that year a Perth (Scone) - Prestwick - Gatwick and a Perth - Newcastle - Gatwick service were introduced, but due to lack of support the services were severely pruned by July 1962.

A line-up at Gatwick! Freighter G-APLH is in the foreground whilst Ambassador G-AMAE and an unidentified aircraft are parked beyond.
(DAS/GMS Collection)

The earliest located timetable - for the winter of 1960/61. The services were all flown by the DH Dove and the fares are interesting: for example, Bristol to Newcastle return cost £10 12s 0d!

DAN-AIR timetable

Winter
1960/1961
(Valid from 1st January, 1961)

A FUTURE AT GATWICK

Maurene Thompson poses with other Dan-Air staff for yet another publicity photograph - this time the launch to the public of the Ambassador. Interestingly, on the original print a cloudscape is visible through the cabin windows, so it must have been taken in flight! (DAS/GMS Collection)

Dan-Air's second international scheduled service opened in January 1962 when a Dakota service was operated between Liverpool and Rotterdam. During the same year the airline flew its Bristol to Basle service via Bournemouth and scheduled services also opened to Ostend from Bristol and Gatwick.

A further Ambassador, G-ALFR, was obtained from Napiers Ltd, the aero-engine company, who had been using it as an engine test-bed. By 1961 their tests were complete and it changed hands, moving to Lasham, where the airline converted it on a long term basis, the aircraft not entering service until January 1964. The fleet was again enlarged in April 1962, when G-ALZN, an ex-Overseas Aviation aircraft, was obtained.

The long hauls continue...
The Australian long-distance Bristol Freighter trips continued to operate, and Captain David Charlwood recalls some of the flights and the people that flew them...

> "I only flew with the Company from January 1960 to September 1961 but I remember at that time we had some marvellous characters flying with Dan-Air. We had a number of full-time pilots that flew all the year round - people like George Sutton, Danny Daniel, Charlie Chaldicott, Denis Bamber, George Fairclough, Pat Falconer, Miro Sar, Dick Davis, Pete Dibley and Jim Torry. Then there were the 'freelances' that only worked for us during the summer season: John Ryder, Ken Coop, Peter Nock and Johnny Devlin...

DAN-AIR SERVICES

G-AMLL, the Freighter that David Charlwood flew in as First Officer, at a rather empty Gatwick. This aircraft was later sold to Lambair of Canada. (DAS/GMS Collection).

On the long flights down to Australia and Singapore we operated with two complete crews, slipping part way down the route. My log-books record a couple of memorable flights that demonstrate just what it was like.

Flight number 2216 (which we took over from Capt. Larkman) was in Bristol Freighter G-AMLL under the command of Captain Danny Daniel with myself as First Officer. I think we had 'Sherry' Sherwood as Navigator, Jack Horan as Engineer and a Radio Operator. The sectors I flew on were as follows:

DATE	FROM	TO	FLYING TIME
25/5/60	Butterworth	Rangoon	5 hrs 05 mins
	Rangoon	Calcutta	4 hrs 30 mins
29/5/60	Calcutta	Karachi	9 hrs 55 mins
30/5/60	Karachi	Masirah	4 hrs 10 mins
31/5/60	Masirah	Aden	6 hrs 25 mins
01/6/60	Aden	Wadi Halfa	7 hrs 10 mins
	Wadi Halfa	Benina	7 hrs 10 mins
02/6/60	Benina	Lyon	8 hrs 10 mins
03/6/60	Lyon	Hurn	3 hrs 05 mins
	Hurn	Lasham	0 hrs 45 mins

I then had six days off before flying on the same Freighter back, this time as Flight Number 2220 under the command of Captain Pat Falconer with Victor Ramchandran as Navigator, Bill Edgar as Radio Officer and Jack Horan as Engineer.

10/6/60	Lasham	Hurn	0 hrs 35 mins
11/6/60	Hurn	Nice	4 hrs 10 mins
12/6/60	Nice	Benina	6 hrs 15 mins
13/6/60	Benina	Wadi Halfa	6 hrs 25 mins
	Wadi Halfa	Aden	6 hrs 55 mins
14/6/60	Aden	Masirah	6 hrs 45 mins
16/6/60	Masirah	Bombay	6 hrs 05 mins
	Bombay	Calcutta	6 hrs 20 mins

As can be seen, some days we flew up to 14 hours - this was before maximum flight hours!. I have many fond memories of this period of my flying career, but to tell some would require permission from my Captains!

By now, the airline was truly international, if only when one considered the nationalities that made up the flight-deck crews! Dan-Air had Australians, New Zealanders, South Africans, Amerian, Dutch, Greeks, Norwegians, Canadian, French, Iranian, Tasmanian, Irish, Swiss, Scottish, Rhodesian, Polish; even Chinese!

A FUTURE AT GATWICK

More Aircraft.
A DH Heron was obtained in May 1963, and replaced the Doves on the Plymouth - Cardiff - Bristol - Liverpool - Newcastle service on 18 September. Early that year the Yorks had been replaced by BEA's own new Argosies on their domestic services, so most of the Yorks were retired by the end of the year, although one machine was kept on hand for long distance charter work. The aircraft's eight-ton payload and large cargo doors had made it an ideal cargo-carrying aircraft, but it also provided Dan-Air with a big problem in terms of finding a suitable replacement. The last machine was retired early in 1964.

By 1963 almost all the Inclusive Tour flights were being flown by the Ambassadors. From Gatwick they served Amsterdam, Barcelona, Basle, Madrid, Milan, Munich, Nice, Palma, Perpignan, Pisa, Tarbes, Zagreb, and Zürich. In addition the Ambassadors appeared in the Manchester IT Saturday flights to Basle and Ostend, and on the Monday flight to Tarbes. The Dakotas were relegated to the Gatwick - Ostend and Rotterdam charters and all the scheduled services to the West Country and Isle of Man.

The airline carried 115,215 passengers during 1963 on all its routes, of which 35,735 were carried on scheduled services. Dan-Air's overall load factor on these scheduled services was almost 43%, although the international scheduled services did little to help, contributing only 1,993 passengers with a load factor of just 19%. The 33,742 passengers carried on the domestic services produced a load factor of a little over 50%.

The Inclusive Tour flights produced a totally different picture - a load factor of 90 percent, mainly due to the excellent passenger appeal of the high-wing Ambassador, which carried 70,402 passengers of the 1963 total. Cargo work was undertaken by the Freighter fleet, with these aircraft being increasingly used to transport racehorses around Europe.

Of sit-downs and slush...
1963 was a year that Captain Arthur Larkman has good reasons to remember:

> "On Good Friday 1963 we were chartered to fly 55 British passengers to Dusseldorf and return with 55 Germans. After landing, the crew carried out their preparations for the return flight and I went to the Traffic Office to complete the paperwork. Keith Moody, the co-pilot - who subsequently became a Type Rating and Instrument Rating Examiner, Ambassador Fleet Manager and eventually 727 Fleet Manager - was always a very calm and solid character. I was surprised therefore, when he hurried into the office in a state of some agitation and said *"They're throwing our passengers back onto the aircraft!"*
>
> We went out to the aircraft where an astonishing sight met our eyes. German police were carrying our passengers, by their arms and legs, up the steps and literally throwing them onto the aircraft.
>
> Urgent enquiries revealed the reason. Our passengers had been in the Customs area when a very large crowd of people in the terminal started shouting a welcome. It transpired that the Campaign for Nuclear Disarmament (CND) and associated

DAN-AIR SERVICES

Ambassador G-ALZY taxies in to the Passenger Terminal at Manchester's Ringway Airport. (DAS/GMS Collection)

organisations in Europe had planned an Easter march and rally in Dusseldorf. Many thousands from all over Europe were taking part and a group had come to greet our passengers, who were all leading members of Britain's CND.

The German authorities, fearing a demonstration, decided to bar entry for our passengers and told them to return to the aircraft. In accordance with CND procedure, they promptly sat on the floor and refused to move, so they were then forcibly removed from the terminal and 'loaded' upon the aircraft! The Chief Immigration Officer flatly refused to change his mind and I was instructed to return to England immediately.

The passengers were very upset by this decision and threatened that if I attempted to take off they would do everything they could to prevent it. A state of deadlock continued into the night, despite appeals to various officials (including the Mayor) and the passengers remained on the aircraft.

At 0730 the next morning we prepared for take-off having warned the passengers of the danger if they did not remain in their seats. However, as soon as we started to roll, most of them left their seats and I was forced to abort the take-off. Once again, we were unable to disembark the passengers on instructions that had by now been confirmed by the Bonn Government, so negotiations continued throughout the day with the added complication that the world's press descended on the airport and dogged our every move.

During the afternoon the rally organisers announced their intention to divert to the Airport for their demonstration. The tension increased and troops were brought to the airport whilst we were told to move the aircraft (complete with passengers aboard) to a hangar. The aircraft was surrounded by armed guards in the hangar and we were horrified to learn that no ground power would be provided and no food or drink would be allowed on board! Conditions on board were very bad, although we were able to open the wing exits to provide some air circulation. After further negotiation, we were able to have the toilets emptied and food and water brought aboard. The stewardesses, Meg Martin and Val Evans, assisted by Engineer Officer Jock Mills, behaved magnificently throughout this trying time, attending to both passenger and aircraft needs.

By Sunday morning everyone was exhausted, so we attempted another departure in order to comply with the authorities' demands that we left with our passengers... Their determination was as strong as ever however, and once again we had to abandon the attempt. We spent the morning in further negotiation to break the deadlock. The break-though finally came in the afternoon when they were persuaded that a large press contingent awating them at Gatwick might drift away if they did not return that evening. At last they agreed to go - we went!"

Towards the end of the year Captain Larkman found himself involved in work that required totally different skills. Instead of negotiating between intransigent governments and unruly passengers he (and the airline) were working on potentially dangerous tests aimed at improving airline safety worldwide.

A FUTURE AT GATWICK

There had been a great deal of controversy about the cause of the crash of BEA Amabassador G-ALZU at Munich on Thursday 6 February 1958 with the Manchester United football team aboard. The German accident enquiry blamed ice on the wings, but this finding was strongly disputed by those who thought that slush on the runway was the principal cause. What became clear was that although there was some knowledge about the effects of aquaplaning in reducing braking efficiency, very little was known about the retardation effects of slush on aircraft take-off performance.

In 1963 the Royal Aircraft Establishment commenced further research at their airfield at Bedford and Dan-Air expressed a willingness to participate in furthering the industry's knowledge of runway contamination to improve flight safety. Thus the RAE chartered a Dan-Air Ambassador and crew for these trials; Captain Larkman takes up the story:

> They planned to use tons of crushed ice spread over the runway to various depths, but as the weather was too cold when the experiements began, the ice would not melt to slush. The 'solution' to the problem was a typical piece of British ingenuity. The RAE scientists discovered that the effect of standing water on the aircraft's performance was identical to that of slush. They therefore constructed three long tanks along the runway, spaced to accommodate the nose and mainwheels of our Ambassador. These tanks were formed by gluing strips of thick rubber to the runway surface so that they would contain water to a depth of about four inches.
>
> I flew G-ALZN to Bedford on 11 November and we commenced a long series of test runs at various depths of water and at various speeds up to 20 knots above V2, which was the speed the Munich aircraft achieved before abandoning its take-off. Photographs were taken from all angles to record the spray patterns of each run and their impingement onto various areas of the aircraft. Rates of acceleration and other parameters were also recorded. These runs could be 'interesting' to say the least - there were times when we aquaplaned so much that we could not stop and therefore had to fly off
>
> These tests continued right throughout November and up to 6 December. From the results gained came a new understanding of the problems set by contaminated runways, which soon became the basis of operational procedures for all types of aircraft worldwide".

A dramatic photograph of Ambassador G-ALZN with Captain Arthur Larkman at the controls during one of the slush tests at RAE Bedford. The tanks containing the water can be clearly seen on the runway in front of the aircraft.
(A Larkman Collection)

DAN-AIR SERVICES

The Ambassador had firmly established itself as the mainstay of the fleet and the 1964 IT programme saw departures from Gatwick to all the usual destinations. The programme from Manchester had been expanded with Ambassador departures on Friday to Munich, Saturday to Bordeaux and Perpignan, Sunday to Basle, Dinard and Ostend and to Tarbes on Monday. The Dakota also flew the Manchester - Ostend Inclusive Tour flight.

The airline had now completed ten years of highly successful flying business and few changes had been made to the airline's management structure. The Directors were F.E.F. Newman, J. W. Davies and L.E.Moore with Captain Warren L.Wilson acting as Operations Manager.

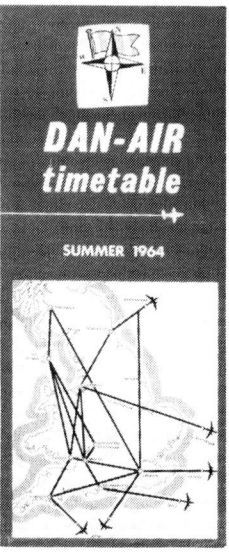

Day charters...

During the mid-1960s the Company became aware of what could be a quite lucrative market - a series of 'extended pleasure flights' - single day charters organised by other companies' Sports & Social Clubs. One such charterer was Ted Hurst, who explains how his particular series of charters came about:

The summer 1964 timetable. (11 April - 31 Oct).

"I worked for Molins Limited (the cigarette machine manufacturers) in Deptford, East London and had heard from my brother-in law about a day out to Jersey his firm had organised. Talking it over with my works colleagues, I got an interested response, so the next weekend I went down to Gatwick to try to get more information on how to arrange it. By chance, the first information desk I saw had 'Dan-Air' above it, so I asked the girl sitting there for help. She simply referred me to their Head Office.

I soon found that there was more to hiring a passenger carrying aircraft than a motor coach - I became an aircraft charterer! I had a contract between myself and the operator, I had to acquire a Board of Trade licence to fly the aircraft over the sector, arrange the take-off and landing times, then to fit the aircraft in with other Dan-Air movements and agree a contract price!. Mr R A Pigeon, Dan-Air's Commercial Manager at Bilbao House, guided me through the process and on 10

Left: The Molins party inside G-AMSU during the flight to Jersey. Ted Hurst is in the centre, wearing glasses. (Photo T Hurst)

Right: The 'Duplicate Original' charter document signed by Ted Hurst and by R.A. Pigeon on behalf of Dan-Air.

40

A FUTURE AT GATWICK

DUPLICATE ORIGINAL

AIRCRAFT CHARTER PARTY

BETWEEN

DAN-AIR SERVICES LIMITED

Bilbao House, 36/38, New Broad Street, London, E.C.2

(hereinafter, and in the conditions hereafter referred to, described as " the Carrier ")

AND

Name MR. E. HURST.,

Address 92, GROVELANDS ROAD, ST. PAULS CRAY, ORPINGTON, KENT.

(hereinafter, and in the said conditions, described as " the Charterer ")

The Carrier will charter to the Charterer and the Charterer will take on charter the aircraft described in the schedule below (hereinafter, and in the said conditions, described as " the aircraft ") for the flight, journey, service or period and upon the terms specified in the said schedule—subject to the conditions set out on the back hereof—which the Charterer hereby agrees and accepts.

AS WITNESS :

FOR AND ON BEHALF OF THE CHARTERER   FOR AND ON BEHALF OF THE CARRIER

No. 8358   THE SCHEDULE ABOVE REFERRED TO   Date 6TH DECEMBER, 1965

(a) AIRCRAFT DAKOTA.   Maximum Permissible Payload 6000 LBS.

Seating UP TO 36 PASSENGERS PROVIDING PAYLOAD STATED ABOVE IS NOT EXCEEDED.

Cargo

(b) Charter Price £ 220—0—0d. LUMP SUM EXCLUDING ALL ROAD TRANSPORT CHARGES.

To be paid to
DAN-AIR SERVICES LIMITED
TWENTY ONE DAYS before commencement of Flight.

(c) DEMURRAGE £ 50—0—0d.   Per Day or *pro rata* for Part of a Day

(d) DETAILS OF FLIGHT FROM GATWICK AIRPORT TO JERSEY AIRPORT AND RETURN.
ON 1ST OCTOBER, 1966.
SCHEDULE LOCAL TIMES.

09.00 – GATWICK – JERSEY – 1025.
2030 – JERSEY – GATWICK – 2155.

N.B.—Passengers will report 45 minutes prior to scheduled time of take-off, and all freight will be presented .... hours prior to each scheduled time of take-off.

DAN-AIR SERVICES

the process and on 10 December 1965 we signed an agreement for 36 passengers to fly Gatwick - Jersey return on 1 October 1966. The total price was £220, with a deposit of £50.

Come the day, 36 of 'the lads'from Molins met up at Gatwick and climbed aboard Dakota G-AMSU for a 9 am take-off. The two pilots appeared to be young men - the Captain wore dark glasses, a crumpled cap and a roll-neck sweater and looked like something from a Humphrey Bogart movie!

We rolled down the runway and after we got airborne, we started to climb in what seemed a series of giant 'steps' to 6,000 feet. The stewardess busied herself bringing round coffee and biscuits, but the boys were interested in a somewhat stronger beverage.

The landing at Jersey (scheduled for 10.25 am) was somewhat dramatic, the aircraft bouncing three times before coming to a standstill to the derisive cheers of all aboard! The return flight was even noisier, as the lads had spent a good amount of time in the Jersey pubs and, clutching our Duty Frees (which we were not allowed) the journey back to Gatwick passed quickly with a sing-song and a few more drinks. The landing at Gatwick at around 10 that evening - accompanied by rousing cheers from those aboard - was perfect and we asked the Stewardess to tell the Captain he was getting better, which she did. On return she told us "I am not going to tell you his reply..." which brought more cheers!

Other aircraft were chartered to other groups for much the same purpose, all of them making use of the Affinity Group Charter rules. Alan Reynolds, who used to organise outings for The Newgate Club...

"During the 1960s I organised a whole series of day trips using Dan-Air. We took an Ambassador full to Paris (flying Gatwick - Le Bourget), the Epernay Vineyards, hosted by Count J. R. Chandon-Moet and the German Wine Cellars around Frankfurt at the invitation of the Export Manager of C.A. Kupferberg & Co. Looking back through my old newspaper cuttings, it seems we only went where there was liquid refreshment available, but what you have to remember is that travel 'abroad' - even on a day-trip - was much more unusual, special and exciting than it is today! Flying by scheduled services then was very expensive and so the charters I organised via Dan-Air meant that the prices could be reduced to affordable levels. We all had a great time, even if the Staff did have to suffer an amount of good humoured banter on the way back!"

During 1965 a Tees-Side - Chester - Cardiff Dakota service was opened, but did not prove successful and so it was dropped after just 27 days! A Cardiff - Bristol - Amsterdam service was also introduced, but this too was soon dropped. Dan-Air's scheduled service to Plymouth also came to an end that year, and the Heron was sold. The summer season saw the operation of a service from Gatwick to Jersey, Gatwick to Kristiansand (via Newcastle) and Gatwick to Ostend, from Bristol and Cardiff to the Isle of Man, Liverpool and Newcastle, from Liverpool to Rotterdam and from Prestwick to the Isle of Man.

Dan-Air's Bristol Freighter fleet had never really been fully utilised, so eventually they were pensioned off, the last two finding their way over the Atlantic to Canada, for Lambair of Manitoba.

One of the Bristol Freighters was exchanged in February 1963 with Handley Page Aircraft as part payment for two more Ambassadors, G-ALZO and 'LZY. Both these aircraft had previously been with the Royal Jordanian Air Force, 'LZO being

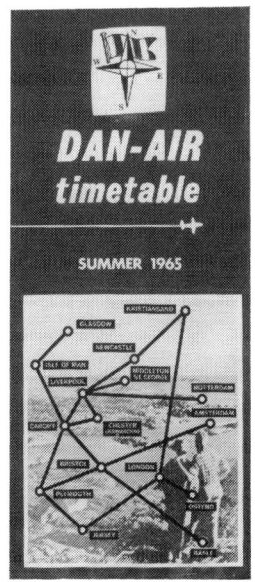

A pair of timetables from the mid-1960s. Above is the summer 1965 example showing the network, including Middleton St George that was served the West Country services. Below is the summer 1966 (4 April - 30 Sept) example.

A FUTURE AT GATWICK

Who do you think you are, Stirling Moss?
One interesting series of charters that many staff wanted to fly was when the airline carried the British Formula One Racing cars and their drivers around Europe. 'Haggis', as usual got herself on the flights...

> "The cars used to go in a Freighter and the drivers in an Ambassador, with possibly yet another Ambassador following behind with rich sponsors, owners or fans aboard. I got to know drivers like Ron Flockhart, Dan Guerney and Sirling Moss very well. Dan Guerney's wife was always nervous of flying, so I got her to help me out in the galley to take her mind off it. The drivers always tried to hide the tension that always seemed to be present before a race by racing miniature racing car models up and down the cabin floor. I often had to step over them to serve drinks!
>
> I remember Stirling Moss returning on one flight almost totally encased in bandages. He was badly burned and bruised with some of his teeth knocked out after an accident somewhere, but he was still able to laugh and joke about it in his usual fashion."

Today Maurene lives near Guildford and has a very different career she's a research and development cosmetic chemist with designs on becoming an authoress as well! After leaving Dan-Air she flew freelance for a while with a number of other airlines before starting a family. Her eldest daughter Andrea later continued in the 'family tradition' during the mid-1980s when she also flew as cabin staff with Dan-Air!

The Bristol Freighter was used for all sorts of loads. Here Wally Baron loads parts of the British Grand Prix racing team's cars, which resulted... (DAS/GMS Collection).

DAN-AIR SERVICES

...in a Freighter full of racing cars!. Visible (just) are five cars. This picture give a good idea of the load carrying capability of Dan-Air's Bristol Freighters. (DAS/GMS Collection).

More Douglas products.

On 4 October 1965 Ambassador G-ALZY opened a Liverpool to Amsterdam service, the Liverpool/Rotterdam service having been dropped. Two Douglas DC-4s - G-APID and 'PXJ - were obtained on lease from Trans World Leasing for use on both passenger and freight flights in place of the Yorks. In addition to using them on IT charters from Gatwick, these DC-4s appeared on the Isle of Man to Prestwick run when loads were heavy.

The DC-4 was a four engined, tricycle undercarriage development of the DC-3 Dakota, powered by four 1,540 Pratt and Whitney Twin Wasp radial engines. These machines were returned to their owner at the end of 1965 and an all-cargo DC-7B/F purchased in exchange. This aircraft took on a role similar to that of the Yorks some three years earlier, operating from Gatwick on long-range charters to Europe, Africa, the Middle and Far East.

The DC-7B/F was the ultimate development of the DC-4, powered by four 3,400 horsepower Wright turbo-compound radial engines, with increased wingspan and fuselage stretch, a taller fin and increased fuel tankage that gave the design a true trans-Atlantic capability.

Dan-Air's example was a large capacity aircraft, capable of uplifting 15-16 tons of freight - twice that of the York - over a

A FUTURE AT GATWICK

*Gatwick in the 1960s - and not a pure jet in sight! On the ramp are a pair of Dan-Air Ambassadors, a Viscount and assorted Douglas products.
(DAS/GMS Collection)*

distance of 2,800 miles at an average speed of 300 mph. The main forward door measured eight feet by seven feet and the main rear door measured ten feet four inches by six feet six inches, which made the loading and stowage of bulky cargo a simple matter.

The aircraft was soon placed in extensive use: flood relief to Italy, a computer to Moscow, emergency relief for a Persian earthquake, equipment for an irrigation scheme in India, even film equipment and vehicles to Tokyo for *'You Only Live Twice'*, one of the 'James Bond' films! Much of the work entailed the carriage of bloodstock - thoroughbred racehorses and prize-winning cattle - including one trip on which Don Conway of the Commercial Department accompanied nine racehorses valued at three quarters of a million pounds belonging to trainer Mr Vincent O'Brien from Milan to Shannon airport, the arrival of which was shown on Irish television, thus generating much free publicity!

Can you fly a 'Dak' Mr Greene?...
Dan-Air not only chartered aircraft and crews out to tour operators but they also very occasionally 'dry chartered' (leased out machines to other concerns to fly using their own crews). One such charter occurred in the autumn of 1965 when G-AMSS was hired by Hughie Green and Rediffusion Television to fly prizewinners of the 'Sentimental Journey' segment of T.V.'s massively popular *'Double Your Money'* quiz-show around Europe!

It seems that when a bell rang during the show, the contestant had won the chance to make a 'Sentimental Journey' to any destination he or she wished. 'Double Your Money' host Hughie Greene (who had flown Dakotas during the war) explains...

> The first contestants to win were two Chelsea Pensioners who wanted to visit the restaurant that was once run by the 'Mademoiselle from Armentieres' featured in the song. There were others; one had a daughter who was a maid in the Vatican, another wanted to see Monte Carlo in the winter and yet another wanted to see the apes at Gibraltar...
> I convinced Redifusion (the company that made 'Double your Money') that it would be cheaper for us to charter a machine and I would fly it myself to carry the

DAN-AIR SERVICES

contestants, film-crew and equipment rather than use the scheduled airlines. We contacted Dan-Air, who gave us a good price, and they agreed to remove the 'Dan-Air London' titles on the side of the fuselage and substitute with 'Double Your Money' , but they wanted to know that I could fly a Dak - surprising, considering I had over 1,000 hours on the type!. Anyway, I did a ferry flight from Gatwick to Luton as a one hour check-ride with Dan-Air's Captain Larkman on the 29th October and the next day for some reason we went up to Manchester and back.

Finally on the 30th we set off from Luton, going first to Lille to film the pensioners. The 1st November saw us going from Lille to Basle then on the 3rd Basle to Rome for the Vatican maid. On the 4th we went Rome to Nice where we stopped for three days then onto Madrid and Gibraltar on the 7th. I flew 'MSS locally around the rock on the 8th before setting off for Bordeaux where we night-stopped, getting back to Luton on the 9th. The next day I delivered the Dak back to Dan-Air at Gatwick undamaged!'

G-AMSS wearing the titles of Hughie Green's TV game show 'Double Your Money'. It is thought the photograph was taken in Rome (Friends of the DC-3)

The Ambassadors continued to give good service. One duty detail that many crews enjoyed was a five day rotation when aircraft flew Gatwick - Athens - Rhodes - Crete. Every five days an aircraft set off to fly to Athens to drop off 55 passengers. Here they would uplift another 55 and fly to Rhodes where the passengers and crew would stay for five days. A new crew that had been enjoying a break in the sun would take-over the aircraft and fly another 55 passengers to Crete for a further change of passengers to return to the United Kingdom.

Ambassador G-ALZX was badly damaged on landing at Beauvais due to poor weather conditions and the machine never flew again. This incident, and the withdrawal from use in 1966 of G-ALFR, created vacancies in the Ambassador fleet which were filled with the timely retirement of the pair of Shell Aviation machines G-AMAA and 'MAG, both equipped with luxury ten-seat interiors. 'MAA was not converted, and was withdrawn from service and used as spares for the rest of the fleet. 'MAG was converted, however, into a 64 seater and put to good use.

The airline was on a sound footing: business was good, with passengers and freight figures on the increase. Never an airline to stay still for long, Dan-Air was now ready to move on to bigger and even faster things..

CHAPTER FOUR
Comet!

During October 1966 Dan-Air signed a contract with the British Overseas Airways Corporation (BOAC) for the purchase of two of the latter's De Havilland Comet Mk.4s. These aircraft had, in fact, been delivered to Lasham some five months earlier, but did not enter service until the winter of 1966, for there was a great deal of work to be done to convert them into 99 seat, high density Inclusive Tour machines. The main cabin floors were strengthened, the seating capacity increased with the fitment of single class interiors and the wings strengthened to withstand the rigours of what would be relatively short-haul flying. The purchase of these two aircraft heralded a whole new era for the Company, for thus Dan-Air became only the second independent airline in the United Kingdom to operate pure jet aircraft, British United being the first with their new Vickers VC-10s. These two machines pioneered the use of jet equipment for the now expanding inclusive tour market and were chartered to Messrs. Clarkson and Horizon Holidays.

The DH106 Comet had originally been designed in great secrecy at the manufacturer's Hatfield factory during the closing stages of the Second World War. Initially planned as a tailless mail-plane for use on trans-Atlantic services, the concept evolved into a more conventional aircraft.

The first flight of the world's first pure jet transport, powered by four 4,450 lb static thrust De Havilland Ghost engines, occurred at Hatfield on 27 July 1949 when the first prototype G-ALVG took to the air in the hands of De Havilland's test pilot John Cunningham. The first public demonstration occurred at Farnborough the same year, with orders quickly coming from many airlines. British Overseas Airways Corporation, Canadian Pacific Airlines, Air France, and the Royal Canadian Air Force all wanted the type.

Sadly, a series of disasters struck the Comet. The prototype was lost during take-off from Rome during October 1952, and CF-CUN was lost at Karachi, after which it was discovered that the pilots had held the nose too high on take-off. G-ALYV was lost at Calcutta following elevator spar failure. But the worst was yet to come. G-ALYP was lost after take-off from Rome on 10 January 1954 followed by G-ALYY at the same location on 8 April. The type was voluntarily grounded for safety checks.

Intense investigation of wreckage and tests on other examples of the type eventually revealed that the problem began with a tiny structural failure of the aircraft's skin, which resulted in severe explosive decompression.

Valuable lessons had been learnt, but the massive lead De

DAN-AIR SERVICES

An advance publicity photo taken just before the first Comets were obtained. Here a Dan-Air stewardess stands with Captains Bob Atkins, Arthur Larkman and Senior First Officer Jock Mills in front of a BOAC Comet at Heathrow. The stewardess's outfit is pale blue with a dark blue hat and white gloves. (DAS/GMS Collection)

Havilland's had built up over its American rivals, Boeing and Douglas, had sadly been lost. The Comet Is had come to an enforced end of their lives, but later examples were completely re-designed structurally and thus evolved into some of the strongest and, in comparison to their size - with four 10,500 lb static thrust Rolls Royce Avon engines - some of the most powerful airliners ever built. The comparatively small number of seventy machines produced went on to give sterling service with many airlines, and in particular Dan-Air.

Further, more and faster.
The acquisition of the ex-BOAC Comets meant that Dan-Air was now capable of flying further afield much more quickly: indeed the utilisation figures for 1967 clearly reflected this. During the year the airline carried 45,643 passengers and some 220 tons of freight, an overall load factor of almost 41 percent. The ITs were flown mainly by Comets, with eight thousand flying hours amassed.

In May new routes were opened from the Isle of Man to Carlisle and Swansea using Dakotas and during October an additional Comet was added to the fleet. From Bristol, the airline operated a Saturday afternoon charter to Ostend on behalf of Hourmont Travel, flown for several seasons by the Ambassadors.

COMET!

The classic lines of a Comet in flight. Here is G-APDK, one of the first obtained, seen above the clouds in the early colour scheme.
(Winged Memories)

The publicity that use of the Comet could achieve was considerable. In May 1967 Dan-Air Comet G-APDK landed at Glasgow to the cheers of over 30,000 football fans welcoming home members of the Glasgow Celtic football team that had just won the European Cup. The team had flown out to Lisbon earlier that week in the same machine, and following the result Intercontinental Caterers were contacted with orders to quickly bake a celebratory cake to be consumed with glasses of Champagne on the flight home!.

Little expansion was possible during the 1968 season, due to the ruling Labour Government placing severe restrictions on the amount of money that people could spend abroad, which in turn severely restricted the public from taking foreign holidays and therefore reducing the demand for IT flights.

In order to make better utilization of the Comet fleet, Dan-Air started flying Inclusive Tours from West Berlin to the Mediterranean on behalf of the West German tour operator Neckerman und Reisen. The first flight occurred on 31 March 1968 when a Dan-Air Comet lifted off from Tegel Airport bound for Malaga. Almost 300 flights were made that summer, making the basing of one aircraft permanently in the city a necessity.

The operation of this aircraft from the city involved positioning a Comet from Gatwick to Berlin early on Monday morning, and returning to Gatwick late on the following Friday. Initially this was an entirely British operation, with staff operating on short stays away from Gatwick. Gradually the Company began to recruit local cabin and ground staff until almost the entire operation (apart from the flight-deck crews) was run by Berliners.

This was followed in November 1968 by the spectacular demise of Harold Bamberg's much-troubled British Eagle, which meant that Lunn-Poly and Everyman Travel were left without aircraft to fly their holiday services during 1969. Dan-Air was quick to react, and immediately tendered to operate these services. Early in 1969, following the award of these two contracts, two former American Airlines BAC 1-11s and two more BOAC Comet 4s were obtained.

DAN-AIR SERVICES

Heading for the sun...
Some of the activities of the time in which the airline became involved may seem slightly strange now, but at the time air travel was still very much a novelty for many of the general public. This interest could not be better demonstrated than when Clarksons, Gold Case Travel and the Newcastle Evening Gazette newspaper organised a 'Holiday Spectacular' for one mid-January evening at Tees-Side Airport. The organisers had expected about two thousand people to turn up, but when the night arrived, over fifteen thousand potential passengers queued, often in driving rain showers, to take a look inside a Dan-Air Comet 4 and a 1-11 that were parked on the airport apron!

Peter and Nola Wilde took one of the early Dan-Air Comets on a five night Clarkson 'City Break'.

> "We went to escape the English winter at the end of March, after the birth of our second child.
> When we arrived over Venice the pilot apologised to us for orbiting the city, but he was awaiting clearance to land as a 'light aeroplane's whereabouts was uncertain'. We were sitting on the 'wing-down' side of the cabin and the view below us was breathtaking! The pilot apologised again, but we shall be eternally grateful for that unknown Italian pilot for giving us such an extended view of what is without doubt the most beautiful city in the world. I've always felt well disposed to Dan-Air ever since those lovely five days, and we could hardly blame them for the snowstorm which greeted us at Gatwick on our return!"

A 'Chorus line' of six girls walk away from Comet G-APDD. The date is around 1969 and by now the uniform has changed - it is now a deeper blue with a white blouse. Skirt lengths have risen too!
(DAS/GMS Collection)

COMET!

Gatwick in 1970 - G-APDK & 'DP await the next load of holidaymakers whilst in the background can be seen 'alterations as usual' being conducted to the airfield.
(DAS/GMS Collection)

1968 also saw a number of Comet crews heading for the sun - but not for leisure! Captain Larkman explains:

"Over the years Dan-Air has operated many aircraft for other airlines, but possibly none was more remarkable than when in July 1968 we were approached by Kuwait Airways to fly their entire Comet fleet whilst they sent all of their own flight-deck crews over to the USA to train on the Boeing 707 (a number of which they were obtaining to replace their own Comets).

A lot of preparation went into this, and we were in close contact with them for about a year before, so eventually we were contracted to operate all their services from October '68 and detached a number of flight-deck crews out to Kuwait. Our pilots flew out to the UK, through the Middle East to India and so on - for a number of months until their own crews returned with the 707s.

It was quite a thing (and possibly unique) for one airline to say to another 'We have the trust in you to take-over our entire operation'. That was a feather in everyone's cap".

A beautiful photograph of G-APDM turning onto the main runway at Gatwick for departure to the sun in the early 1970s.
(Winged Memories)

1969 was a year of rapid expansion. In May the former British Eagle Gatwick to Newquay service was opened by a Dan-Air Dakota, in place of the British Eagle Viscount and BAC 1-11, a flight that now took 75 minutes. Somewhat understandably, passenger resistance to the placement of the much older DC3 type on a route that had been served by more modern jet aircraft meant that Dan-Air did not find this service as successful as had been hoped, so by the end of the season it was dropped.

DAN-AIR SERVICES

In October another 'first' occurred - one of the Comets opened Dan-Air's first ever trans-Atlantic passenger flight when a party of tourists was flown to Port-of-Spain, Trinidad.

G-APDE of the Training Unit, surrounded by ground equipment, is prepared for another flight. Note the unusual 'platform' conversion to bonnet of the Morris Minor pick-up truck of Engineering in the foreground! (DAS/GMS Collection)

Simulated training

As the size of the Comet fleet grew, so did the need for training facilities. A Dan-Air Group training centre was opened in Nightingale Road, Horsham, the brainchild of Deputy Chief Pilot and Chief Training Officer Captain Arthur Larkman. Originally opened in 1967 as a pure ground school, the building was expanded to house the Comet simulator and provided training facilities for Group members and a large number of outside companies.

The Comet simulator, originally manufactured by Redifon at nearby Crawley, was purchased from BOAC and was, at the time, ten years old. Nevertheless, it could be programmed to simulate ten different degrees of difficulty, ranging from normal flight to a full-scale emergency. When the machine was moved from BOAC's Cranebank facility, their simulator engineer moved with it. The complex piece of equipment took twelve weeks to install and check out. Newly-joined pilots had to fly a total of 28 hours on the machine and spend eleven weeks at the centre. After initial training the pilots returned annually for instrument checks.

In spite of the savings in costs and time by using the Comet

COMET!

simulator, such was the demand for aircrew to convert to the jet that Dan-Air felt it necessary to set aside one machine - Delta Echo, under the control of Captain Tom Walters - to use as a dedicated training aircraft. Although mainly based at Tees-Side, the aircraft visited many other UK and European airports. To give an idea of the number of pilots converting to the Comet, during 1971 the number of crews increased from 28 to 43. Altogether 22 would-be Captains and 28 would-be First Officers undertook the course, putting in a total of just under 4,000 hours! Just a year later Dan-Air had over one hundred crews flying the type!

Apart from the Comet aircraft, Dan-Air purchased the spares holding stock from BOAC, which would have had an original value of well over £1,000,000. This meant that Dan-Air could not only provide spares for other operators, but also provide the engineers at Luton, Manchester, Glasgow and Berlin with a comprehensive package of essential high usage items so that normal maintenance could be carried out at the airports concerned with the minimum of disruption to Comet operations.

It was not surprising that with all this expansion 1969 was the first year in which the airline carried over half a million passengers - a total of 509,025 passengers on Inclusive Tour and scheduled services. 29,867 were carried on the scheduled routes and 469,158 on the Inclusive Tour and other charter services.

We meet every eventuality...
Introduction of the Comet into Dan-Air service proceeded smoothly, both crews and aircraft settling down into the routine, although not every flight was straightforward, as the voyage report for Comet G-APDP on a trip to Tunis from Gatwick on 26 June 1969 demonstrates. Captain Martin, First Officer Duthie and

An unidentified Comet in the early colour scheme undergoes maintenance at Lasham. Note the myriad of equipment and scaffolding needed to conduct a major overhaul. (DAS/GMS Collection)

DAN-AIR SERVICES

Engineering Officer Casey were on the flight-deck with four cabin crew looking after a full load of passengers...

The flight was normal until entering the Tunis Terminal Manoeuvring Area (TMA) when the Dan-Air crew were advised that a light aircraft had just called an emergency, lost and with three minutes fuel remaining. The Captain contacted the aircraft in difficulties on the VHF radio band and found that he was about 40 nautical miles south-east of Tunis. The Dan-Air Captain led him to a disused aerodrome where the light aircraft pilot carried out a successful forced landing, the Dan-Air pilot passing his position to Tunis Air Traffic Control.

The report goes on to state somewhat laconically that "*Departed Tunis eleven minutes behind schedule due (to the) above, arrived Gatwick on time*". It seems that Dan-Air flight crews were ever helpful, and not only to their customers!

There was another flight in which the passengers suffered a unscheduled stop-over at an airport they never thought they would ever visit! The occasion was a Comet flight from Berlin's Tegel Airport to Istanbul in Turkey under the command of Captain Incledon-Webber (known to all as 'Inky') and his crew. This was in the early 1970s, the days of 'corridor flying' over Iron Curtain countries which imposed severe restrictions on where a Captain could take his aircraft. A flight plan had been filed that allowed the aircraft to take the shortest route, across Yugoslavia and then over Bulgaria. Inky takes up the story...

"Air Traffic Control let us into Yugoslavia and we were just about to hand over to the next when the previous ATC controller came on the radio and told us to get out of there quick - which we did. We were then resigned to going all the way around the border. It was not long before ATC were on the radio again '*Sorry about that, that was a mistake - do come back in*'. Well we did (because it was so much shorter) and resumed our previous course. Not long after that I found myself flying in very close formation with a pair of MiG jet fighters - one on each wing-tip, escorting me in for a landing at Sofia in Bulgaria!"

*An unusual view of one of Dan-Air's Comets on the ramp at Gatwick, surrounded by much of the ground equipment needed to prepare it for another flight.
(DAS/GMS Collection)*

COMET!

*The ex-Channel Airways G-APMB with the undercarriage almost up climbs away from Manchester Airport. By now the colour scheme has changed - a stronger red line runs under the cabin windows and the tail is all-red with the 'Compass points & Flag' logo in a white circle.
(DAS/GMS Collection)*

Captain Incledon-Webber joined Dan-Air in 1969, and has a fund of hilarious stories, many of which are totally unprintable. One that is (and which ably demonstrates his wry sense of humour) shows how rumours get around and can get out of control! Although 'Inky' started the story as a joke, eventually everyone in the industry swore it was true and in fact, many swore they knew someone who had seen it!

The setting was the Holiday Inn, Bristol, where 'Inky' and his crew arrived one evening for a normal overnight stop-over. In the reception area was a huge model of *'Concorde'*, which was being built at Filton near Bristol at the time and everyone locally was very proud of the fact. All the crew admired the model upon arrival and speculated whether Dan-Air were ever going to get any.

Inky and one of his cabin staff were coming down for breakfast the next morning when they noticed that overnight a wing had been broken off the model. The girl remarked what a shame it was. 'Inky' agreed and, keeping a straight face, told her the Management were charging him £300 to put it right. "*Why, did you break it?*" asked the girl. Inky replied '*affirmative*' (even though he had been nowhere near it). "*After you went to bed last night the rest of us sat around talking and someone bet me that I could not hang-glide it off that balcony*". He indicated a balcony two floors up above the reception desk. The girl looked at him in amazement "*You jumped off there?*" "*Sure, it was no problem, and I would have made it as well. I was at the right angle of attack, just coming into land when, because the wings of the model had not been stressed, it broke. A real shame that...*" He looked ruefully at the model.

'Inky' thought the girl had realised that he was gently pulling her leg, but for ages afterwards he was known as 'the man that had hang-glided a model of Concorde'. Eventually the 'joke' wore thin, so he decided to enlighten everyone, only to find that no-one believed him. "*Oh no Inky, I know it's true*" was one reply he got, when trying to put the record straight. "*...I know someone who shares a flat with someone that is a friend of someone who saw you do it!*"

DAN-AIR SERVICES

G-APDM rests on the stand at Gatwick. The Pinnion fuel tanks on the wings are particularly clear in this view.
(DAS/GMS Collection)

A 'Case' history!

Passenger care is important for all airlines, and Dan-Air is no exception to this; nevertheless, there are instances when it seems that a passenger almost insists on not being helped!

On a Dan-Air flight returning from Palma, Majorca, one unclaimed suitcase was left in the customs hall. The airline was immediately notified by the airport authorities and wrote to the London address on the suitcase label to explain what had happened. Nothing was heard for a month or so, so the airline wrote again and there still was no acknowledgement. Dan-Air obtained release of the case from Customs and Excise after they had removed all the duty-free goods, and on opening the case the Company found Nursing Certificates issued in New Zealand and an address. A further letter was despatched to New Zealand explaining what had happened, and from there the information was forwarded to the owner of the case in England, who was extremely pleased to collect it, some three months and a lot of hard work after she had lost it!

If there was a problem with the Comet it related to passenger baggage, for the under-floor cargo hold was not very large. However, by placing a weight restriction on the amount of luggage carried, it was adequate for holiday-makers. This was a minor problem that was more than offset by the ability of the aircraft to carry a useful number of passengers around the holiday haunts of Europe and, with its power reserve, operate out of many of the smaller hot airfields without any load penalty.

'Casevac' from Africa...

Bruce Fardell remembers one episode when the Comet's performance was required to the fullest extent...

> "One day I got a call that I had to be at Gatwick within two hours to take a Comet with a full medevac team aboard out to Abidjan on a mercy flight. Some Costain construction workers had been injured there in a car crash. First I had to find out where I was going and then work out the route. Abidjan was a tiny little airport near

COMET!

Lake Chad, so the route was out to Malta and down across the Sahara. The charts told us that they had a runway, but on arrival we discovered that it was not in commission, so we had to do a visual flypast to work out what was going on.

Eventually we got down and rigged up a lorry to get the stretcher cases aboard. The strip was tight, even for a lightly loaded Comet, so first we measured its length with the mileometer on a borrowed Land Rover, then paced it on foot before working it out on the charts as to how much fuel we could uplift. The figures told us we could not make it back direct to Malta, so we would have to call into Kano; by the way, all of this went on in temperatures of around 35 degrees Celsius!

We had to 'hurry up' the Africans to get a move on so we could get off and away before those on board started to 'cook' in the heat! The girls meanwhile were searching through their flight-bags for anything they could use to make them as cool and comfortable as possible!"

Sky Diners...

A further 'first' for Dan-Air occurred when they became the first airline world-wide to introduce disposable catering equipment aboard their aircraft. Known as the 'Sky-Diner', the concept was developed by E.S.& A Robinson Ltd as a result of a suggestion by Mr J.R. Blake of Intercontinental Air Caterers (Gatwick) Ltd, the Dan-Air caterers, in order to reduce the weight problem on certain aircraft and routes.

'Sky Diners' consisted of a larder made from a special board, coated with metal foil to minimise fire risk, into which fitted 18 meal or 36 snack trays, also made from special board. The chief advantages of this equipment were the savings in weight, time and improved hygiene standards, for no cleaning was necessary.

On 11 June a party of travel agents, tour operators, airline executives and members of the press were taken from Gatwick to Bristol in order to introduce the 'Sky Diner', which was generally well received. At the time of the launch the Company announced that use of this equipment would achieve a direct saving of around 4d (2p) per passenger meal - not much per passenger, but with half a million passengers flown during 1969, that could be over £10,000 annually!

'Juliet Kilo' (G-ARJK, a Comet 4B) also in the interim Comet colour scheme, rests at Gatwick between flights with flaps down. (DAS/GMS Collection)

DAN-AIR SERVICES

The same year, the Birmingham based consortium Midland Air Tour Operators introduced a series of two-week holidays from Birmingham's Elmdon Airport to the Greek islands of Rhodes and Corfu using Dan-Air Comets. Prices ranged from what today seems an incredible £73 for a two week stay in the Messongi Beach hotel on Corfu during the low season to £108 for the equivalent holiday at the Coreyra Beach. So that everything would proceed smoothly, four days before the first departure the airline flew out a party of 100 travel agents for an 'educational visit', a quite common occurrence during the late 1960s and early 1970s when any new destination was about to be served. One presumes that in the case of Greece, it was to explain the intricacies (and hazards) of drinking the local Ouzo to the agents so that they could pass it on to their customers!

The 1970 Inclusive Tour season saw a fleet of twelve Comet 4s in service, operating mainly from Gatwick and Manchester. From Gatwick the airline operated to Alicante, Athens, Djerba, Dubrovnik, Faro, Gerona, Ibiza, Ismir, Malaga, Palma, Rhodes, Rimini, Sardinia, Tenerife, Tunis and Varna. The Comets flying from Glasgow served Alicante, Gerona, Ibiza, Palma and Rimini. From Manchester they flew holiday-makers to Alicante, Gerona, Ibiza, Palma, Rimini and Venice. Edinburgh also was the start of a weekly Comet service to Palma.

Sadly, this was the year that tragedy struck the airline. On 4 July 1970 Comet 4 G-APDN, en-route from Manchester to Gerona with 107 Clarkson holiday-makers and seven crew flew into the side of a mountain near its journey's end, killing all on board.

This accident came just two days after Dan-Air announced that it had signed a four year deal with Global worth £2.5 million which covered the operation of all Global's flights from Birmingham from April 1971 onwards.

More and more Comets...
Two more Comet 4Cs were purchased in March 1971 and were placed on some of the longer-distance European charters in a 119-seat configuration. Charters operated to Las Palmas, Tenerife and

The ultimate Dan-Air Comet colour scheme! G-BDIX, an ex-RAF C.Mk.4. The area around the cockpit glazing and above the red cheat line is now black and the Company name has dissapeared from the wing fuel pods. (DAS/GMS Collection)

COMET!

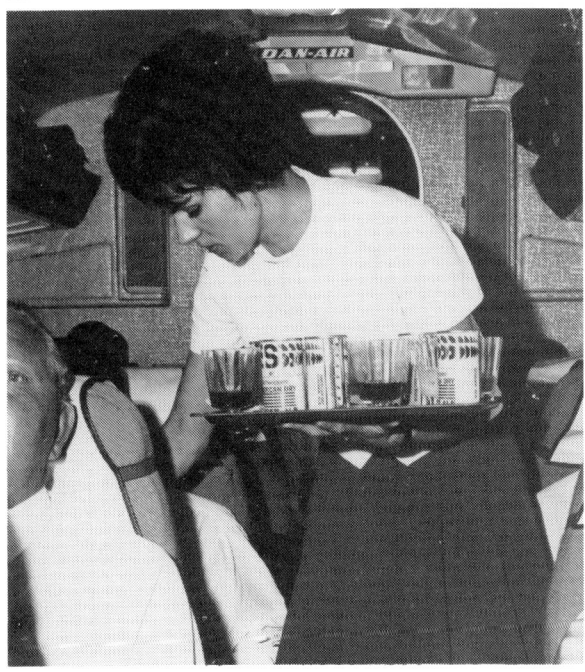

"Would you like a drink Sir?" The working conditions inside a Comet, around 1972. Note the lack of overhead bins, the apparent darkness of the bulkhead behind the stewardess and the closeness of the seats. How things have changed! (DAS/GMS Collection)

many of the Greek islands, which were just becoming very popular with British holiday-makers.

A number of BEA Comet 4Bs had been transferred during the early 1970s to their newly formed associate company BEA Airtours for non-scheduled Inclusive Tour work. Airtours gave the Comet only two years of economic life and thus had already started looking for a suitable replacement. The Comets were progressively replaced during 1973 by a number of Boeing 707s, allowing Dan-Air to obtain them. The last BEA Airtour Comet flight occurred on 31 October 1973 and the remainder of the 4B fleet soon found its way to Lasham for storage and re-painting for use by Dan-Air as required.

With the turn of the decade, Gatwick, the airline's main operating base, had undergone a transformation. For a number of years since its re-opening many people in the airline and travel industry regarded the airport as a 'white elephant' but by 1972 it ranked as the tenth largest in Europe and had one of the largest growth rates in the world. The airport was in the throes of a massive expansion programme with a planned handling of 8 million passengers by 1975. 1974 saw the third runway extension completed, giving about 10,000ft of available concrete.

Although the Comet found favour with many thousands of passengers, there were times when Dan-Air's high-density seating configuration of the main cabin caused problems. Peter Neighbour recalls possibly the most uncomfortable flight he ever had...

"In the mid-1970s I was employed by Occidental Oil and regularly commuted between Gatwick and Tripoli. The Company always used British Caledonian Airways, for it was possible to leave Houston, Texas one evening, arrive at Gatwick at 09.00 the next morning, catch the 10.00 flight to Tripoli and be there in the early afternoon.

Unfortunately, there was all the cargo door problems with the DC10 and every one was grounded. B.Cal maintained their trans-Atlantic services using 707s but were compelled to charter-in replacement aircraft to use on the Tripoli run. The aircraft used were Dan-Air Comet IVs which had been used on holiday flights. Whilst they were doubtless adequate for this, they were totally unsuitable for the carriage of large Texans from Gatwick to Tripoli, especially after a long trans-Atlantic flight! Several guys were physically incapable of getting into the seats and I vividly remember some of the (unrepeatable) comments prior to the removal of the armrests.

I genuinely felt sorry for the crew, who strove valiantly to make everyone happy, but in some cases, I am sorry to say, they were unable to succeed. Luckily the DC10s were soon back in service, otherwise I had visions of mass lynchings at Occidental's Travel Department!

To be fair, this was the only occasion on which my experiences of flying Dan-Air has been less than perfect. I currently fly regularly from Nice to Gatwick and, at present (1992) Dan-Air is well ahead of the pack, largely because of the consistently high standard of course, and a very good underlying sense of normality and humour".

Twilight of the Comet

By the mid-1970s the Comet was now beginning to reach the end of its useful life. Nevertheless, it says much for the skills and expertise of the engineers at Lasham, for they obtained a contract from the Ministry of Defence to partially modify one of the fleet, G-APDP, into a test-bed aircraft for the Nimrod Airborne Early Warning (AEW) machine. This aircraft, now serialled XX944, was delivered to the Royal Aircraft Establishment at Bedford during June 1973.

For a number of years the airline's engineers had experienced problems in the location and acquisition of spares to keep their Comets flying, so Dan-Air scoured the world for any examples they could obtain as spares or replacement aircraft for their fleet. In 1975 the airline acquired a number of low-hour Comet C.Mk.4s from the Ministry of Defence that were in excellent condition. They had been used until its disbandment on 30 June 1975 by 216 Squadron, Air Support Command of the Royal Air Force.

These aircraft were brought up to civil standards by the engineers at Lasham and quickly placed into service along with three ex-Egyptair machines, allowing some of the earlier aircraft to be retired. The problems of finding ways of keeping the Comet fleet airborne still existed, so further examples were obtained from East African Airways, Aerolineas Argentinas, Middle East Airways and others. Most of these aircraft were flown into Lasham, where they languished for a while, donating items every now and then to keep the others flying. Eventually, almost all were broken up.

With the so-called fuel crisis it could be expected that the Comet would be retired in favour of more economic aircraft, but with the book value written down to practically zero and adequate spares backing available, the type still retained its appeal to the airline. It says much for the passenger appeal of the Comet to the public - and the part the aircraft played in the story of the airline - that one source quotes that during the 1979 IT season, a massive 30% of all British holiday-makers used Dan-Air Comets!

Out to grass...

Not all the examples of the world's first jet airliner were scheduled to end their days on the scrap-heap, as a number of Dan-Air Comets managed to escape the breaker's torch intact. Many of Dan-Air's Comet fleet had interesting histories, but one in particular deserved

COMET!

preservation. G-APDB, which went to the East Anglian Aviation Society (later the Duxford Aviation Society) for their growing civil airliner collection at the Imperial War Museum's outstation at Duxford near Cambridge, held the honour of being the first ever fare-paying passenger jet airliner to cross the Atlantic, which it did on 4 October 1958 whilst in service with BOAC. Another aircraft, G-APDK, remained at Lasham as an unusual clubhouse for the Lasham Air Scouts. Eventually, this machine was also scrapped.

1979 finally saw the demise of the Comet when Dan-Air's last surviving 4B, G-APYD (ex-Olympic and Channel Airways) touched down at Gatwick on a flight from Heraklion, Crete on 23 October. One week later 'Yankee Delta' took off for what was to be its final flight - in the hands of Captain Joe Wright - for the aircraft had been donated by Dan-Air to the Science Museum's Civil Airliner Collection at their outstation at Wroughton near Swindon for preservation. When the aircraft touched down at its new home it was its 18,586th landing and it had completed some 32,738 flying hours - or some fifteen million miles!

I must be on that flight!

The last commercial flight of a Comet was made on Sunday 9 November 1980. The aircraft was 4C G-BDIW, had been specially chartered by John Hunt of Ian Allan Ltd to fly a group of aviation enthusiasts determined to mark the type's passing in style, for this was to be not only the airline's, but the worlds last commercial Comet operation. Before moving to Ian Allan, John had been Reservations Manager with Dan-Air and was determined that the type operated by his old employers was to go out in a blaze of glory. Dan-Air was happy to oblige. DA8874 - as the flight was listed - was officially a one hour roundabout flight to Bournemouth and back. However, the flight departed Gatwick in the hands of Captain John Kelly heading for low fly-bys at London Heathrow, Brize Norton and Lyneham before returning to Gatwick!

119 passengers boarded, with others waiting seven hours in vain

After its last flight on 12 November 1979 G-BDIV is put out to grass at Lasham to await its fate. (DAS/GMS Collection)

for a standby seat! Two passengers who had booked seats but arrived late from the Midlands due to traffic delays threatened John Hunt with physical violence if he did not let them aboard!

Assisting Capt. John Kelly were Capt. Simon Searle and Flight Engineer Gordon Moores, a highly experienced Comet crew, the two Captains having flown 31 and 34 different examples respectively. Looking after the passengers were Fleet Stewardess Val Barnett and her enthusiastic cabin staff.

When viewed today, the interior of these machines looks archaic - ancient almost, so rapid has been the pace of aeronautical progress. Lacking are the overhead luggage 'bins', the sculped ceilings and discreet lighting in the main cabin. The flight-deck has seating for five, but was normally flown with four occupants - Captain, First Officer, Flight Engineer and Navigator. Nevertheless, many thousands of holiday-makers made their first-ever flights to the sunspots of Europe and further afield in Dan-Air's Comets and the preservation of a number of machines ensures that the important part the design played in the development of both the airline's and the world's aeronautical heritage is available for future generations to study.

Thus the era of Dan-Air's Comets passed, but a commercial concern cannot become sentimental. The Comet design had come to a premature twilight of its years as a fuel thirsty and expensive-to-operate jet, so it was time again to consider the future. With the demise, someone within the Company calculated that during the fleet's life, the legendary Comet had flown a total of some 238,000 hours in Dan-Air service, which equated to a distance of 95,400,000 miles!

No chapter on the Comet would be complete without including one photograph that demonstrates its sparkling performance! Fittingly taken aboard the last commercial flight, the wing angle to horizon shows 'IW going up like a homesick angel!
(John Hunt Collection)

DAN-AIR SERVICES

CHAPTER FIVE
Modernisation

Dan-Air had become well established in the first ten years of it's operation, but the civil aviation field is ever-changing. The era of large piston engined aircraft that flew at relatively low speeds and altitudes was coming to and end and the entire fleet of Dan-Air needed to be replaced; but with what? Mr F. Newman provides and insight.

"The year 1964 was a most critical one, because it had become necessary to decide whether to go forward or give up. The era of piston engined aircraft was ending and the step up to turbo-props and jets was a major one. The industry was tightly controlled and opportunities to break out of the licence stranglehold that gave BOAC and BEA a near monopoly on scheduled services was very limited.

However, not without some dissension, the decision was made by Dan-Air to go on. Having taken the decision, what should be the next step? It was obvious that the fleet needed to be modernised, but BAC 1-11s were expensive and difficult to obtain. The problem was solved when we purchased two Comet 4s from BOAC, which our engineering Division converted into 99 seaters.

From now on for the next 20 years the inclusive tour business became the backbone of the Company's commercial business. During this time, Mr Alan Snudden became Managing Director and was well supported by Mr O'Reegan from Cooper Brothers, who became Finance Director.

The Commercial Department was expanded my Mr Tapling who was involved in the trans-Atlantic charters with the 707 fleet and the operation out of Berlin, where four aircraft were fully employed. Mr Siddaway also joined and was responsible for the Inbound Charters from Europe and assisted in the Inclusive Tour market, where Mr Cossey supported the Managing Director".

N5041 (later G AXCP) one of a pair of BAC 1-11s obtained from American Airlines. This aircraft (seen here at Lasham undergoing conversion to UK standards) was to remain in service with Dan-Air until 1986. (DAS/GMS Collection)

As already mentioned, Dan-Air managed to obtain a pair of BAC 1-11s in 1969 from American Airlines for the Lunn-Poly and Everyman Travel contract. The 1-11 was designed as a pure-jet successor to the superlative Vickers Viscount turbo-prop of the 1950s. Its origins can be found in a design project study by Hunting Aircraft Ltd of Luton, who had been working on plans for a small jet airliner from as early as 1956, identified on paper as the Hunting H107.

DAN-AIR SERVICES

With the take-over of Hunting by the British Aircraft Corporation (BAC) in 1960, work proceeded on the H107 as a joint project to a point where in May 1961, British United Airways placed an order for ten machines, the design having grown to seat a maximum of 79 passengers in a high density cabin layout.

A range of variants was built, offering different layouts, weights and ranges to suit customers requirements. Powered by a pair of rear-mounted Rolls Royce Spey engines that gave a cruising speed of 550 mph, the early variants of the 1-11 carried 89 passengers and at that time it was one of the world's finest short-haul airliners.

Dan-Air's first two BAC 1-11s were only two years old when they were obtained from American Airlines, but as with many of the aircraft acquired by the airline, they required a tremendous amount of work before they could enter service. This was not only to meet the airline's own specifications, but more importantly they required 'Anglicising' meaning that all the fuel indicators had to be changed from pounds to kilograms and a considerable amount of other flight instrumentation in the flight-deck had to be altered to meet British Airworthiness Registration Board requirements.

The galley unit also had to be re-engineered to suit the airline's operations, the toilet moved from the front of the cabin to the rear and the passenger cabin converted from 76 to 89 seats.

Time was very much of the essence in order to meet the schedule imposed by the Lunn-Poly contract, and the probelm was aggravated when, soon after the contract to purchase the aircraft was signed, and before an independent maintenance check could be carried out on the machines, the American Airlines mechanics went on strike! Nevertheless, the first BAC 1-11 arrived at Lasham on 14 March and had its United Kingdom Certificate of Airworthiness (C of A) issued just 19 days later, which says much for the dedication of the Dan-Air engineers!

Bases at Luton and Manchester

The Lunn-Poly contract was particularly large, the two 1-11s being almost fully occupied during 1969 on these services alone. It was also the year that brought about big changes in both cabin and ground staff uniforms, for Dan-Air changed from the traditional blue uniforms to a more easily recognisable crimson; the skirt lengths also reached higher altitudes!

For the contract, aircraft were based at Luton to fly holiday-makers to numerous Mediterranean resorts,

The new uniform for 1969 seen here modelled by Ground Receptionist Val Rodway. Out went the blue outfit with white flouncy blouses, in came a flame red two-piece with 'Mandarin' collar top.
(DAS/GMS Collection).

MODERNISATION

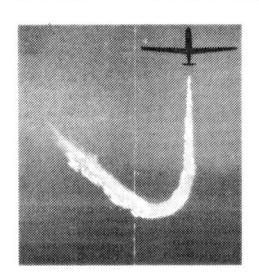

The earliest located In-Flight brocure, thought to be from around 1967.

Dan-Air's winter timetable for 1966/7.

including Alghero, Alicante, Djerba, Gerona, Ibiza, Istanbul, Luxemburg, Mahón, Malaga, Munich, Palma, Pisa, Rimini, Santiago, Split, Tunis and Venice.

This meant the rapid formation and development of a new Dan-Air base at the busy Luton Airport, located to the north of London. Stuart Harbottle was chosen as Base Manager, with Captain Dickie Draper as 1-11 Fleet Captain, Eddie Brown as Base Engineer (later replaced by John Fox) the Senior Hostess being Janet Ford.

With three duty officers, 23 BAC 1-11 pilots, 5 Comet crews, 50 hostesses and 12 engineers based at Luton, and only 500 square feet of space allocated, facilities and office accommodation were, to say the least, cramped, but they all formed themselves into an efficient team in a very short time. Three offices (for traffic control, load control and the duty officer) were constructed, with three passenger reception desks allocated in the main airport concourse.

At the summer peak, some 60 flights a week were leaving Luton on the Lunn-Poly contract, meaning that the airport was soon closely rivalling Gatwick for the number of Dan-Air flights. By 1971 Gerry Penwarden (latterly Director, Customer Services) was Base Manager of Dan-Air Luton. He can well recall one chaotic mid-week in June:

> "The European Cup Final - between Parathanaikos of Greece and Ajax of Holland - was due to be played on Wednesday 2nd June at Wembley. A large amount of support was expected from both teams and a huge airlift had been organised to get their fans over to see the match.
>
> Luton was ideally situated to accept flights from both countries and transfer passengers to Wembley, slightly to the north of central London - that's why Dan-Air Luton received a Telex message from Amsterdam requesting scheduling approval for 45 aircraft turn rounds. Owing to the large number of movements involved and the restrictions on night jet flights, a staff member from Dan-Air London would visit Holland; from the discussions that resulted, Transavia (the Dutch charter airline) agreed to send across a team of engineers and a member of their traffic staff to assist at Luton. At the same time, scheduling approval was finally gained for 25 of the flights requested, to be operated over a period of three days using Caravelle and Boeing 707 aircraft.
>
> Before the onslaught began, it was necessary to streamline our procedures somewhat in order to accept such a large number of foreign passengers in such a short space of time. To do this British landing cards and visitors cards were despatched to Amsterdam with instructions that all passengers should complete these forms before landing. Extra staff 'volunteered' for duty and extra ground equipment was brought up by road from Gatwick.
>
> Finally the aircraft started to arrive. There were periods of intense activity followed by monotony whilst the aircraft returned empty to Amsterdam to operate another trip. In the Arrivals hall, hundreds of singing fans packed tightly together, trying to abduct Dan-Air ground receptionists. There was drama when two plane loads of Greek supporters came face to face with three plane loads of Dutch fans. Luckily nothing was thrown - not even the lone Security man separating them!
>
> After the match, with the terminal building packed with victorious Ajax supporters and the bar that stayed open until 5 a.m., only confused impressions remain. A duty officer who took 25 minutes to cross the 25 yards between the two Dan-Air offices. Ramp crews trying to con Ajax baseball caps for the supporters while the check-in girls fought off wave after wave of good humoured attacks. Aircraft after aircraft arrived and departed like clockwork.
>
> Finally they were all gone - the tired, drunk and lame - even the virtuoso with a trumpet, leaving comparative peace in the terminal at last."

DAN-AIR SERVICES

The airline had been flying into Manchester since the very early days, but with the inclusion of the airport into the scheduled services network, it became clear that it was to play an increasingly important role in the Company's future. Nearly 100,000 passengers were being carried from there, mainly on Inclusive Tours, with flights totalling 55 per week in the peak summer months. Two Comets were permanently based at Manchester and were supplemented by a further Comet and a 1-11 at the weekends, so there was the need to establish both an Reservations and Operations facility there.

The Senior Operations Officer was Vic Jones, with Duty Officers Ian Lambie, Edgar Leporta and David Stead. Reservations were handled by Christine Brooks and Angela McGrath.

Flight crews could (and did) find themselves doing the strangest things, often well outside their job descriptions, as the Trip Record Sheet for Flight No DA1083 on 26 June 1969 shows. The aircraft was G-AXCK, one of the new BAC 1-11s that had just entered service. It was being flown by Captain Mitchell and First Officer Champreys with three stewardesses, C. Watkins, R. Warren and M. Dunn looking after the passengers on a flight to Venice. The report reads..."*Satisfactory flight outbound. Venice still on strike. We (the stewardesses) cleaned cabin and toilets ourselves. Captain and First Officer loaded baggage. Therefore we were delayed. Otherwise, everything O.K.*"

For the 1970 Inclusive Tour season from Luton, the 1-11s flew to Alghero, Alicante, Ibiza, Dubrovnik, Gerona, Rome, Luxemburg, Mahón, Malaga, Munich, Palma, Rimini, Santiago, Split and Venice.

No chance to publicise the airline was missed! Here three of the Manchester-based stewardesses, (l. to r.) Beryl Bazley, Ann Butterworth and Anne Close, visit the set of Granada Television's world-famous 'Coronation Street'. (DAS/GMS Collection).

MODERNISATION

One million carried, but the 'Dak' soldiers on...

Despite the ever rising operating costs, 1971 was to be another successful year for Dan-Air, with its expansion plans continuing to develop. For the first time ever, the airline carried over one million passengers, the large majority of whom were on its Inclusive Tour operations on behalf of British tour operators.

The late autumn, winter and spring saw a dramatic increase in people taking short, cheap holidays and this contributed much to higher aircraft utilisation during those traditionally slack months.

Even in the mid-1960s mass tourist air travel was very much a novelty. In those days a trip to Palma was regarded as long distance, and very up-market, so travellers 'dressed up' in their Sunday best for the flight. The so-called 'bulb-flights' - single day trips to Holland to view the tulip fields - were very good business for Dan-Air and did much to popularise air travel for pleasure. The flights usually arrived early in the morning at Rotterdam and very soon as many as thirty airliners (up to ten of which could be Dan-Air) would descend on this relatively under-used airport from all over the United Kingdom. Almost all would be on charter to the tour operator Clarksons, so it is not surprising that this daily invasion each spring became known as the arrival of 'Clarkson's Air Force!

It was not only the passengers that had a different attitude: some of the charterers looked upon their customers differently as well and offered a style that sadly, has all but disappeared. For example, on some flights, each lady passenger received an orchid as she boarded! All of this meant that travel by air for pleasure had a certain amount of 'style' and this could be reflected in the behavour of the passengers; they were polite to the cabin staff and there are many instances recorded where Dan-Air aircraft arrived at its destination with as clean a main cabin as when it departed. This situation was greatly different to some litter-strewn aisles that are left when the passengers de-plane from todays jets!

J.W. Thompson was on one of these day-trips, organised during April 1966 by a friend to the Dutch tulip fields from Bristol Airport. The memories of this flight conjure up a different era of flying...

> "We flew into Rotterdam aboard a DC-3. The aircraft was all carpeted out and in the cabin were small, leather-bladed fans for air-conditioning. We were cleared for take-off and taxied out to the end of the runway, then we came back to the apron. The pilot walked through the cabin to the rear door, saying "I won't be a minute, I've forgotten my sandwiches'. As it was he secured a panel under the wing and we took off. The stewardess was setting up a bar on a table at the rear of the aircraft during which she managed to drop a bottle of gin. The girl was clearly upset about it, for apparently she had to make up the cost from her own pocket! We had a collection for her as a reward for her excellent service which I am sure more than compensated for any expenses she had to find! By the way, we went from Rotterdam to Kaukenhof via Delft by coach and then back to the airport via the Hague - all for an inclusive price of £9.00!

The summer timetable for 1967. (1 April - 30 Sept)

The winter timetable, effective from 1 January 1968.

DAN-AIR SERVICES

We have had many other flights with Dan-Air and on one such flight my wife and I struck up a conversation with one of the stewardesses. She told us her father had piloted DC-3s and Ambassadors and her mother had also been a Dan-Air stewardess, so it seems that Dan-Air is a real family Company!"

The airline may have been modernising the fleet with pure-jet aircraft, but there was still a place for the 'Dak'. Some DC-3s that had been in service with the airline since its inception were still giving sterling service, but during 1969 the airline started looking for a replacement type for its ageing Dakota fleet, which it still flew on the West Country services and the odd charter.

It was during one of these later DC-3 flights that the passengers received a very unusual in-flight announcement. Rosemary Murch, now Mrs Rosemary Cooper, worked in Cabin Services at Gatwick, but at the time was a stewardess on the DC-3 flying out of Bristol. She explains what happened...

"At that time it was 'not the done thing' for members of staff to have 'relationships', so myself and my then boyfriend Roger Cooper (now my husband), who was a First Officer for the Company on DC-3s at the time, took ourselves off to Palma to get engaged. In order to keep it a secret I did not wear my engagement ring to work, keeping it instead on a chain around my neck. Somehow Captain Pat Falconer found out, and on one trip from Bristol to Liverpool I think, with my fiancé in the second seat on the flight-deck, Pat announced over the cabin P.A 'the recent engagement of our co-pilot and air hostess' to all the passengers aboard. I dived to the back of the 'plane to cover my red face!"

Rosemary's husband Roger, latterly on the Boeing 737 fleet, recalls the 'days of the Dak' with much affection:

"On looking back to my early days of flying as a First Officer with Dan-Air, a couple of flights spring to mind. I was involved in the operation of Dan-Air's last DC-3, G-AMPP, from Bristol's Lulsgate Airport, which was a much less restricted environment than today. The first occasion relates to when we set off from Bristol for the short hop across the Channel to Cardiff on the first leg of the scheduled service that called at Cardiff, Liverpool and Newcastle before returning the same way. A fortunate coincidence was that on this day our track across the Bristol Channel was directly over the incoming hulk of the 'SS Great Britain' that was being towed into the Port of Bristol for refurbishment.

Much to the delight of both crew and passengers (most of whom were regulars anyway) we were able to obtain clearance from ATC to make a couple of low orbits above this piece of maritime history as it completed the last leg of its long journey from the Falkland Isles. It is interesting to think that the 'Great Britain' is now almost back to pristine condition, whilst I believe that 'PP is now part of a theme park on the outskirts of Paris!

My other memory of the Dakota concerns the day when the aircraft almost took off without me! We were never on the ground long at Cardiff, so it was quite common for the Captain to start the engines and prepare the aircraft right up to the point of taxi whilst the co-pilot was still closing up outside. The co-pilot was then expected to jump into the aircraft via the rear door and make the somewhat steep climb up to his seat in the cockpit.

On this occasion I was despatched to the Met.

The secretly engaged couple pose on the steps against Dan-Air's DC-3 G-AMPP.
(Mr & Mrs Cooper)

68

MODERNISATION

The summer 1971 timetable, with the Nord 262 on the cover.

Office to update our weather information for Newcastle (where we were expected to arrive in a couple of hours) as it could be quite poor at times. As I started the couple of hundred yards back across the apron, I noticed that the DC-3 had started engines, all ground equipment had been moved away and was starting to taxi out - this time the co-pilot was not struggling up the isle, he was watching it prepare to take off without him!

Fortunately our Traffic Officer, who had been assisting with engine start-up had seen me appear from the Met Office, so he jumped on a nearby motorised three-wheel baggage truck and headed across the apron towards me! On this new, novel form of crew transport I was left standing on a small platform behind the driver, one arm tightly wrapped around his neck, the other holding on to my cap - I tried to remain correctly dressed at all times - as we slowly overtook the taxi-ing DC-3. By this time the Captain, possibly realising that his First Officer was not scrambling up through the passengers and into the cockpit, peered out of his left-hand window to see me formating on his port wingtip astride the tug. The DC-3 was brought to a short stop while I clambered from my smoky, vibrating transport and boarded the aircraft to calmly walk through the passenger cabin as if this was quite normal procedure!"

In order to find a Dakota replacement, a number of machines were investigated, including the Britten-Norman Islander, De Havilland Canada Twin Otter, Beech 99 and Handley Page Jetstream. The latter two emerged as favourites, with the Jetstream demonstrator (G-AXEK) flying the Inter-City service for a few days. An order was placed with Handley Page for a single machine, but the plans were upset when H-P went into receivership in 1970, so once again the airline was searching for a 'DC-3 replacement'.

A number of aircraft were evaluated, including a type 262 test machine from the French manufacturer Nord. An example used as a sales machine was demonstrated to Dan-Air wearing the airline's colours, but carrying the manufacturer's test registration of F-WNDD. In June 1970 an Air Ceylon Nord 262 was obtained.

The Nord had a roomy pressurised cabin with seating for 29 passengers. A major advantage was that only minimal ground equipment was required during turn-rounds, for the aircraft had

Here G-AXEK, the yellow, black and white Jetstream demonstator belonging to CSE Aviation, is seen marked with the titles 'Dan-Air Commuter'. Standing in front are the airlines evaluation team and representatives from Handley Page Aircraft. (DAS/GMS Collection)

DAN-AIR SERVICES

Nord 262 F-WNDD, the company demonstrator painted in the proposed Dan-Air colours. The aircraft still retained its French registration. (Societe Nationale Aerospatiale)

air-stairs built into the passenger door, an electrically operated freight door, and could be started on internal batteries.

The ferry journey from Ceylon to Melun-Villaroche in France (where the aircraft had to be certificated for British use) was of epic proportions - it took fifteen days! On board was a Ceylonese crew flying the aircraft, five engineers and a Dan-Air representative, who had to pay for all the expenses. One small airstrip used to refuel in the Pakistani desert did not appear on any maps. The only way they could find it was to radio for help from a passing Pakistan International Airlines Fokker F-27, who told the 262 pilot to descend to a low level, follow a large sand-dune and where it ended turn right and he would see a white line in the sand! Upon landing, the pilot stopped both engines, only to discover that they had a flat battery and there was no ground power for a re-start! Fortunately another PIA F-27 was due in and was able to bring in a spare battery. After certification the new machine was registered G-AYFR and arrived at its new Newcastle base on 3 July.

Strangely enough, although the Nord was a much more advanced aircraft than the DC-3 that it was to replace on the West Country services, there was an amount of passenger resistance to it

The single Nord 262 G-AYFR is prepared for another scheduled service. (DAS/GMS Collection).

MODERNISATION

upon introduction to the services, Dan-Air received many letters asking for the old type to be brought back. It may well have been that many passengers had become used to the cold, draughty and noisy 'Dak'!

A period was set aside for crew training before the aircraft entered service on the Newcastle - Liverpool - Cardiff - Bristol route on 22 July. The Dakota had finally come to the end of its useful life and the last example - G-AMPP - was finally retired on 22 July after spending the last nine years of its career operating the Bristol - Newcastle service. Dan-Air's last 'Dak' was then flown to Lasham for preservation as G-AMSU, their first aircraft, while the airline's engineering facility at Lulsgate was closed.

On 6 July the Ambassadors opened a weekend service via Carlisle to the Isle of Man. The service was flown throughout the summer, but direct Newcastle - Isle of Man flights were not allowed, for that licence was held by another airline.

Trans-Atlantic Affinity Group Charters

The next stage of Dan-Air's operations was to expand firmly into the trans-Atlantic market. Early in 1970 the British Board of Trade granted an extension of their Air Operator Certificate to cover trans-Atlantic flights, so Dan-Air applied for traffic rights to the USA and appointed the firm of Bridgeman, Long and Pyeatt to act as attorneys for this complex, long-winded task.

As a result of this, an American Civil Aeronautics Board (CAB) application was filed in February, followed by a pre-hearing conference attended by all interested parties. The case was then studied by a CAB examiner at a hearing, who in turn submitted his report to the CAB proper, who then passed its recommendations to the American President for a final decision. Finally, in Washington DC on 29 October 1970, President Nixon granted the airline a permit, initially valid for five years, to fly passenger charter flights into the US from Britain and other parts of Europe.

Dan-Air Services announced that it would commence these operations on 1 April 1971 and that probable equipment would be

G-AYSL, Dan-Air's first Boeing 707, seen at Gatwick during the summer of 1972. (DAS/GMS Collection).

DAN-AIR SERVICES

Douglas DC-8s. A new joint marketing company was formed to help promote this venture - Dan-Air Intercontinental, formed jointly between Dan-Air and CPS Aviation Services.

As it turned out, Dan-Air did not buy DC-8s but on 7 January 1971 G-AYSL, a Boeing 707-321, was handed over to Dan-Air by Pan American at Nassau. The aircraft arrived at Newcastle the next day in the hands of a Pan Am crew to begin a period of crew training before introduction to airline service in March.

The aircraft was mainly used on the trans-Atlantic Affinity Group charter flights from Gatwick on behalf of CPS Jetsave, although during the busy summer months it was used occasionally on Dan-Air's more heavily trafficked European tour routes.

Affinity Group charters had started in the early 1960s, mainly as a result of Caledonian Airways' desire to exploit the strong ethnic links between the big Scottish communities in Canada and North America and their home country. Caledonian decided to make use of a little-known and neglected provision in the travel regulations in order to offer cheap fares. IATA members had always tried to protect their scheduled traffic, but in 1963 they had adopted a resolution which allowed its members to negotiate cut-price charters *'with one person on behalf of a group whose principal aims, purposes and objectives are other than air travel and where the group has sufficient affinity existing prior to the application for charter transportation to distinguish it and set it apart from the general public'* - in other words, not the public in general, and not anyone that intended to fly in any case.

The airlines quickly realised that these links could not only be ethnic, as in Caledonian's case, but also that an 'affinity of interests' embraced a whole range of shared interests as well as blood ties.

The trans-Atlantic charter business boomed, with affinity groups set up on both sides of the Atlantic, fostered by the airlines. It was into this very challenging market that Dan-Air entered with its Boeing 707s. The lure of less than half of the scheduled ticket price for crossing the Atlantic tempted passengers in their thousands, but inevitably the very cheapness of travel eventually corrupted the affinity charter concept. It did not take long for unscrupulous travel organisers to get in on the act, creating fake societies, backdating memberships, and even receiving stolen tickets - anything to get around the regulations. The airlines may have been totally innocent of all this, but nevertheless, they were made responsible for policing the charterers, and watching the airlines to make sure that they did were the Department of Trade in the United Kingdom and Civil Aeronautics Board in the United States of America.

By early 1970 it became clear that all but the closest of the affinity groups existed only to obtain cheap travel, and were being

MODERNISATION

numerous trans-Atlantic passenger charters from Prestwick, Gatwick and Manchester to both the USA and Canada. They were also used on European inclusive tour charters from Gatwick, Glasgow and Manchester at the height of the season.

Dan-Air's first year of operation on trans-Atlantic flying was not as successful as had been hoped, due in the main to strong competition from other airlines. As the Chairman, Mr Fred Newman, said to the shareholders at the 50th AGM of Davies & Newman Holdings on 31st May 1972:

A souvenir booklet and route map 'presented with the compliments of Dan-Air Services Ltd'. From the contents, the period is the early 1970s.

> "The tremendous growth of long-haul charter flights touching a mass market previously uncatered for has posed the problem of how to best licence the business. As 1971 was the last year in which British airlines operated under the joint jurisdiction of the Department of Trade and Industry, the Air Registration Board and the Air Transport Licensing Board there has been considerable uncertainty in the last six months as to how the long-haul charter business should develop. This has been detrimental to some British independent airlines, and in a small way to ourselves. However, guide-lines have been given by the Government to the new Civil Aviation Authority (which started operations on 1st April 1972) and we feel confident that the Authority will interpret these in a manner that will give enterprising operators a real chance to develop".

However, there were other forces at work effecting the trans-Atlantic charter business. In early 1970 it became clear that all but the closest of the affinity groups existed only to obtain cheap travel, and were being widely misused. The airlines continued to struggle against the tide, but it became clear that the majority of the general public were only interested in cheap, mass-travel transport. Affinity travel was so corrupted it became unworkable.

In June 1972 the United States CAB announced that because of the many violations of affinity charters a number of airlines, including Dan-Air, would require advance approval for every charter into the USA. Obviously, it would be impossible for the airline to demand proof of compliance from every passenger! Mr Newman recalls the problems these 'fake' affinity groups could cause:

> "An aircraft would land and the inspectors would go on, look at everyone's affinity groups cards and then declare about 30 as fakes. It was then down to the airline to do something about them; we had to get them back to their departure point."

Something had to be done. New rules relating to trans-Atlantic charters were being drafted, which eventually resulted in the ABC (Advance Booking Charter) [called the TGC (Tour Group Charter) in the USA], but this did little to help anyone but the scheduled carriers. The scheduled airlines were not prepared to lower their scheduled fares, so an alternative was needed and it eventually emerged with Freddie Laker's 'Skytrain'.

The studies done for the purchase of DC-8s for use on the trans-Atlantic services did not go to waste. Captain Arthur Larkman had been involved in these discussions, and recalls how all the work was eventually placed to good use:

DAN-AIR SERVICES

"Air Spain operated a small fleet of ageing Bristol Britannias on charter work. Their owners, The Banco del Nor Oeste wondered, as they were losing a lot of money and were facing bankruptcy, if they could be saved. The bank called Dan-Air in to advise.

After looking at the figures we thought they could, and took over the running of the company. I had been involved in looking at the DC-8 for Dan-Air and thought that it was an ideal aircraft for their needs, so we went out to Eastern Airlines in Miami and concluded the deal getting rid of the Brit and re-equipping with DC-8s. We ran Air Spain for three years until Dan-Air changed into a public company".

Out with the old - in with the new!.

With new aircraft arriving steadily, it became clear that some of the older machines in the fleet were due for retirement. For many summers the sound of the Ambassador's Bristol Centauri had become a familiar sound around many regional airports at home and abroad, but after twelve summers' hard work the type had been gradually phased out. The ex-Shell executive machine, G-AMAG, had been damaged in a wheels-up landing at Manston on 30 September 1968 due to the failure of the starboard emergency uplock release. Captain Moody was flying the aircraft:

"Funnily enough, it was the last day of flying for the aircraft before it was withdrawn from service. We had been doing a number of training details from Gatwick and I landed back there and let most aboard off.

With just Captain Jim Leckie, who was an ex-fighter pilot and later was to become a successful barrister, and myself aboard, we took off again for one last flight. On completion of this, we discovered that the undercarriage would not come down.

We made a number of attempts around Gatwick to try and shake the leg down before we were diverted to Manston (although via the radio the Company engineers wanted us to try and put it down on the grass alongside the runway at Lasham, which I refused to do, for it was too short to make a wheels up landing) We had to circle for ages, to burn up fuel, for the Ambassador had no real fuel dump facilities. We also had to wait for them to foam 3,000 feet of the runway; I believe it was the first time a civil airliner used the emergency facilities there.

We got down safely and, apart from damaging the propeller tips and scraping the belly the aircraft could have been easily repaired. The Accident Investigation people were soon on the scene - they cleared the Company, the aircraft and us of any blame and I was flying again 24 hours later!"

G-AMAG sits amongst the foam on the runway at Manston whilst a Westland Whirlwind emergency helecopter hovers overhead. (K Moody Collection).

MODERNISATION

G-ALZO, the last surviving aircraft plodded on, still giving sterling service, but not without the odd incident!

In its final summer of operation 'Zulu Oscar' was chartered for a number of 'Champagne Days' - £14-a-head day-trips to the Champagne district of France. Towards the end of one such day out in May, sixty merry returning passengers had just boarded the aircraft at Rheims and the Captain was taxying out towards the runway when everything came to a sudden halt. Michael Shaw, one of the passengers aboard, takes up the story:

> "It was hilarious! We'd all had a good day Champagne tasting and had just settled down in the plane ready for the journey home when we were asked by the pilot 'Can you give us a push?' Everybody was still laughing and sparkling from the Champagne as they trooped out of their seats and gathered around the nose to help with a three point turn!"

It seems that Ground Control at Rheims had guided the Ambassador down a blind taxi-way on which it was not wide enough to turn round! The only way out was backwards, so it fell to the passengers to provide a hefty human-powered 'push back' so that they could go home!

The historic last scheduled flight of the Ambassador occurred from a wet Gatwick Airport on Sunday 26 September 1971 when 'Zulu Oscar' lifted off on a normal service to Jersey. On the return, the machine brought back 32 passengers, Company officials and admirers to Gatwick. The very last commercial flight occurred three days later when appropriately, the Dan-Air Social Club took 'ZO on a day out to 'Champagne Country', landing at Rheims. Flown by Captain Pete Collier and First Officer Peter Watts, the passengers were looked after by stewardesses Joy Moore and Judy Jacobs. On return to Gatwick the aircraft was guided into No. 1 Bay to the internal accompaniment of *'Auld Lang Syne'*!

This flight marked the end of twenty years of commercial flying by the Ambassador and in Dan-Air service the fleet had flown well over ten million miles - but it was not to be quite the last! The next weekend 'ZU was dragged from retirement at Lasham to fly a

Farewell to the Ambassador! Members of the Dan-Air Social Club gather for one final official photograph in front of G-ALZO, now preserved and under restoration at Duxford. (DAS/GMS Collection).

75

DAN-AIR SERVICES

replacement 'Spey' engine out to Zagreb where a 1-11 had gone 'tech'.

At the time Frank Horridge, then Dan-Air's Managing Director, was quoted as saying that *"the Ambassador was an outstanding airliner that should have been developed and that, when fitted with propeller turbines, it would have been superior to the Viscount"*.

With the acquisition of the prototype Avro 748 series 2 on 1 May 1971 the airline moved into a new era. Later that month the aircraft moved to Newcastle to start work on the Newcastle - Kristiansand service as the 262 was becoming too small for the loads carried. A second machine was acquired in late 1971, which meant that on 21 January 1972 the last flight was made by the Nord and it was sold to Rousseau Aviation the following month.

The Hawker Siddeley 748 design originated in 1957 as yet another 'Dakota replacement' by A. V. Roe and Co Ltd. Initially a high-wing layout was proposed, but following further research studies, a low wing was adopted, for it offered ease of maintenance and better flying qualities and allowed the fitment of a shorter, sturdier undercarriage to raise the fuselage away from the ground when the aircraft operated from unprepared airstrips.

The Series 1 variant was powered by a pair of 1,600 shaft horsepower Rolls Royce Dart 6 Mk. 54 turbo-shaft engines, while the series 2 version was fitted with Rolls Royce Dart 7 Mk. 531

A well known photo of 748 G-ARAY in the first Dan-Air colour scheme worn by the type. Often thought to be a 'real' air-to-air, this is, in fact, a very clever photographic montage, with the titles overlaid on a picture of 'AY.
(Winged Memories)

Another 'montage', this time a 748 in the colours of Dan-Air Skyways.
(Winged Memories)

MODERNISATION

engines of 1,910 shaft horsepower each. In 1963 the manufacturing company was taken over, and thus the design became the Hawker Siddeley 748 overnight. Later work brought about a further increase of power which gave improved performance and even better operating economics.

Despite the initial adverse reaction from passengers, the single Nord had been reasonably successful. Regularly, operating crews achieved five minute turn-arounds on the scheduled services, an aspect of quick operations that had never been tried before in the United Kingdom. On many occasions this allowed the aircraft to complete its day's schedule well within the allowed time. The Nord also played a small but very important part in the change-over to decimal coinage. Just before 'D-Day', 15 February 1971, as the change-over day was known, the Nord was chartered to carry cheques and credit card slips from Newcastle and Liverpool into Luton to be moved onwards into London by Security Express so that they could be cleared and converted to the new system before the big day!

1972 was to be another year of expansion for the company, with the purchase of Skyways International for £650,000 from Sterling Industrial Securities on 11 February, a deal which was completed by April. Skyways International (previously known as Skyways Coach-Air Ltd) had been around since the early 1950s and had been formed by Eric Rylands to operate coach-aeroplane-coach services between London and the Continent using a small fleet of Douglas Dakotas. The service remained virtually unchanged up to 1961, when the airline hit the headlines by ordering six brand-new Avro 748s - the first order for the type to come from the UK.

Apart from operating the Coach-Air services, Skyways flew a number of IT charter flights around Europe. Skyways Coach Air's associated company, Skyways, was taken over in 1962 by Euravia,

The Dan-Air Skyways timetable for the Summer 1973. (1 April - 31 Oct)

Here seen wearing a later colour scheme is an unidentified 748 parked on the ramp at Bern during turn-round. (DAS/GMS Collection).

DAN-AIR SERVICES

Promoting the company! Above: The 1972 'roadshow' visits Newcastle to advertise both the scheduled routes and charter services.

Below: Dan-Air and Skyways staff prepare to launch the Birmingham to Newcastle service to the media and travel trade. (DAS/GMS Collection).

which later itself changed name to one that is well known by holiday-makers, the Luton-based Britannia Airways. Skyways Coach-Air remained independent and continued operating until early 1971 when Sterling Industrial Securities backed a management buy-out and took over the company, which then commenced trading under the title 'Skyways International'.

From Skyways Dan-Air inherited a fleet of four Avro 748s and, with two of its own and a further machine obtained from BOAC, the company started to build up a new network of scheduled services. The former Skyways machines flew in the colours of Dan-Air Skyways and although they continued to operate the cross-channel

MODERNISATION

coach-air services from Ashford (Lympne) to Beauvais, many of the 748s were transferred to Dan-Air's 'Link-City' network.

Mr. Newman: *"Steps could not be taken until the end of the Summer 1972 season to fully integrate the 'Skyways' operation into our own, but nevertheless, this eventually happened and we began to see the effects on our figures during 1973"*

One person who started his career in 1968 with Skyways as a Commercial Student and is now Commercial Manager, Commercial Relations for Thomas Cook at their Head Office in Peterborough, is Cris Rees. In between, he had quite a varied career with Dan-Air:

"Following the take-over I worked under Archie Lowden, Dan-Air's Reservations Manager. By 1973 I was ready to move onto other things, but was talked into staying by Frank Tapling, and then helped David Langston as 'Assistant Duty Free Manager' - an official title! At the suggestion of Traffic Manager Peter Davies, I applied for and got the job of Station Manager for the Isle of Man, working at Ronaldsway Airport.

Here, I had to be jack-of-all-trades, with just one full time member of staff to help me - a girl that worked Monday to Friday, mainly looking after the Reservations. Apart from the Dan-Air flights to Newcastle, Prestwick, Bristol/Cardiff and Swansea we also used to look after British Midland Viscounts that came in regularly, plus a number of other airlines. When things got really busy - usually at the weekends - I used to hire senior boys from the Public School nearby. I had to drive the ground power truck, do the start-up signals, remove the chocks, look after the catering, fill in the load-sheets, check the passengers in - it was just like having your own little airline in miniature!

During this time the Isle of Man - Swansea service ran only at the weekend, so we had a part time Station Manager there - her 'real' job was working in a bank! This worked out fine until one Sunday we ran very much behind schedule and the aircraft arrived at Swansea on Monday morning with no one there to meet it!

The IoM job was seasonal, so come the autumn I was looking for another post. I went back to the UK as 'Relief Airport Person', working at Speke, Luton and Gatwick, before becoming the Station Manager at Ashford. Dan-Air then moved the base to nearby Lydd, where the high-point of the week was the arrival of the weekly 1-11 service - the local population used to turn out for that!"

The Dan-Air Skyways timetable for the winter of 1972/3. (29 Oct - 17 March)

The French & English Dan-Air Skyways timetable for the winter of 1973/4. (28 Oct - 31 March)

Dan-Air also inherited several licences from Skyways for international scheduled services to Montpellier and Clermont Ferrand, to be flown by Dan-Air from Gatwick.

A new three-times-a-week international scheduled service was opened with G-AZSU on 5 June 1972 when the aircraft flew from Gatwick to Berne, which had an airport too small to accept contemporary jet aircraft, so the service was flown for many years with the 748 fleet.

Filling the seats...

Cris Rees took a break from Dan-Air for a while after the move to Lydd, but soon returned to the fold to work under a charismatic gentleman called Colonel 'Bing' Crosby who ran Romanic Tours, one of Davies & Newman Holdings' subsidiaries. Romanic were at that time running short break tours in an attempt to take up spare capacity on the Dan-Air Scheduled Service routes:

DAN-AIR SERVICES

"We offered tourists a number of destinations - '*Chateaux of the Loire*' and '*Brittany*' based on the Bournemouth to Dinard route, '*Normandy*' based on the Lydd - Beauvais service, '*Provence*' on the Gatwick - Montpellier and numerous others. When Colonel Crosby retired, I managed Romanic and we generated a lot of business to Norway as well"

One regular Dan-Air passenger has very strong reasons for remembering the Beauvais - Lydd operation, and for flying Dan-Air... Miss Jill Goulder from London:

"I always fly Dan-Air if I can, because, many years ago they probably saved my life! On Sunday 3rd March 1974 I flew back from Paris with a friend after a weekend break. When we arrived at the house of friends in London where we were staying overnight, they greeted us with 'we thought you were dead!'

It was the day of the Turkish Airlines DC-10 incident when a ground engineers strike had grounded most planes so most tour groups had been transferred onto the Turkish flight. Dan-Air had its own engineers so was one of the very few other planes (perhaps the only one) that left Paris for the UK that afternoon"

Re-organisation and 'going public'.

The considerable expansion of past years eventually brought about a major re-organisation of the management structure of the Group, culminating in its becoming a public limited company on the London Stock Exchange during 1971. Davies & Newman was changed to Davies and Newman Holdings Ltd, with a new Davies & Newman formed to handle the shipping side of the business. On the aviation side, Dan-Air Services Ltd and a new subsidiary Dan-Air Flying Services Ltd, were formed to handle all operational matters in the same way that Dan-Air Engineering handled all technical matters.

Arrangements were completed by 30 September, resulting in an application for a quotation on the Stock Exchange for the shares in Davies & Newman Holdings Limited. The Company was brought to the market by Hambros Bank with Cazenove & Co. as brokers on terms which capitalised the Group at around five million pounds. 1,133,000 shares went on offer at 130p each, and the issue was over-subscribed.

In general, the financial press were cautious in their appraisal, mainly because there were no comparable companies on the market, and thus no guide-lines. However, it was generally agreed that both the yield and price-earnings ratio (5.8% and 10.5% respectively) were realistic.

Preferential allotment of 52,200 shares was made to employees of the Group. In the prospectus it was revealed that Group profits had risen substantially over the previous few years and that Dan-Air was now contributing around 65% of the total. The decision of the Group to go public marked another milestone in its development and proved to be a fitting introduction into 1972, when Davies & Newman celebrated 50 years of business.

The Chairman and Managing Director of Davies & Newman Holdings Ltd was F.E.F. Newman MC, with J. W. Davies as

MODERNISATION

Boeing 707 G-BEVN undergoes major maintenance at Lasham. Some of these aircraft followed the American practice of carrying the 'last three' numbers of the construction number in small black letters on the nose as an aid to identification. The English convention was to use the 'last two' letters of the registration on the undercarriage doors where possible. (DAS/GMS Collection)

Deputy Chairman and B. M. O'Regan F.C.A. Other Directors were A.H. Langworth, E. J. Mordaunt, H.N. Marten MP and A. J. A. Snudden.

Directly under the control of Dan-Air Services Ltd were Dan-Air Engineering Ltd, Airways Leasing Co. Ltd. and Romanic Trading Ltd. (two aircraft leasing organisations), Dan-Air Intercontinental Ltd (sales agents) and D & C Chauffeur Cars Ltd. Associated companies were Airways Catering, Gatwick Handling (the Gatwick Airport handling agent which was 50 percent owned by Davies and Newman) and Dan-Air Bonded Stores.

The Bonded Store that operated since the mid-1960s in conjunction with Intercontinental Air Caterers. Initially this worked from a small 250 square foot bonded area at Lowfield Heath before moving into Intercontinental Air Caterers' base at Charlwood. In June 1970 Dan-Air Bonded Stores took possession of a new 5,000 square foot warehouse on the Crawley Industrial Estate that held bonded cigarettes, perfume and other Duty-Free goods for themselves and other airlines, plus a number of Duty-Free airport shops. A further bonded store was built at Bristol and the section took over the bonded store at Tees-Side.

Gatwick Handling Ltd. came into being by a somewhat tortuous route. Airlines the world over were known to prefer to process their own passengers, but it is impossible for them to do so at Gatwick, especially when one considers that during 1971 some 60 different airlines used the airport.

For many years Dan-Air had operated their own handling unit in conjunction with third party handlers. During the late 1960s the then concessionaire, Herbert Snowball's Airbourne Aviation, formed a new company with the support of Messrs Melcalfe and Foukes; the name of the new organisation was Gatwick Handling.

DAN-AIR SERVICES

Passengers start to board this unidentified Boeing 707 at Gatwick. (DAS/GMS Collection).

This company quickly went into liquidation and, in order to gain additional check-in desks, Dan-Air with the agreement of the British Airports Authority (BAA) took over the outstanding debts and the name and continued to operate Gatwick Handling.

It is interesting to note that the reasons for continuing to trade under the same name was to save confusion and keep the costs down; travel agents had already had brocures printed advising passengers to report to Gatwick Handling desks for the forthcoming summer season!

With the requirement for even more facilities, Dan-Air entered into talks with Caledonian to form a joint venture. The talks progressed well, but Caledonian changed tack and took over British United Airways. Dan-Air were left looking for a new partner and found it with the newly-formed Laker Airways, who were seeking autonomy at their home base.

Agreement was reached with the BAA and, in February 1972, Freddie Laker, Allan Snudden and David Livingstone (the BAA Airport Director) signed the contract appointing GH as the airport concessionaire. The contract, initially for ten years, made it possible for plans to be made to arrange the large capital investment needed to provide additional equipment to handle the quieter, wide-bodied jets such as the Lockheed Tristar, DC-10 and Boeing 747.

With the demise of Laker Airways, their shareholding of 50% was taken over and split equally between Delta Airlines and Northwest. Each company provided Directors for the Board of Gatwick Handling. For list see Appendix One.

MODERNISATION

Allan Snudden (centre) one of the mainstays of Dan-Air for many years poses with what is thought to be the prototype of a new cabin staff uniform. (DAS/GMS Collection).

Boeing 707 G-AZTG in a colour scheme that had been quickly adapted from the Air Malta livery; the aircraft had just returned from a period of lease. (DAS/GMS Collection).

Strikes, restrictions and problems...

If 1972 was the year of re-organisation, 1973 was the year in which external problems adversely affected the company. A strike by the French air traffic controllers during the spring caused much dislocation of traffic and increased flying times to many destinations, whilst currency problems added to the difficulties to be overcome.

Introduction of the long-awaited Advanced Booking Charters arrived too late for adequate marketing during 1973, but nevertheless, the two Boeing 707s (which were obtained for long-haul charters) were well utilised.

Towards the end of the year supplies of fuel were reduced, causing much time and effort to be spent in finding new techniques to conserve useage and ensure sources of supply. Indeed, the massive fuel price increases brought about through the so called

DAN-AIR SERVICES

'Oil Crisis' of late 1973 and early 1974 caused the airline to switch as many flights as possible from the fuel-thirsty Comets to the more efficient BAC 1-11 and Boeing 727 fleets.

The crisis had been brought about during the closing weeks of 1973 when the Organisation of Petroleum Exporting Countries (OPEC) decided to flex its muscles and use its crude oil resources as a weapon, applied via severe restrictions placed on its supply. This meant that OPEC could control the price of oil for its own benefit and raise the price of what the industrialised world had came to regard as an inexhaustible supply of an inexpensive energy source. In immediate practical terms, this restriction meant that less fuel was available to the airlines, and what there was became rapidly more expensive. After a while the world-wide panic relating to the price and supply of crude oil died down, the price of refined fuel stabilised and the Comets again found favour with the Company.

In August 1973 Dan-Air signed a £1.7 million contract with CPS Jetsave, the British advanced booking charter (ABC) organisation covering the operation of a series of trans-Atlantic ABC flights during 1974. There was to be up to seven flights a week from Gatwick, Manchester and Prestwick to Calgary, Vancouver and Toronto in Canada, carrying up to 54,000 passengers a year. In addition a fortnightly Gatwick to Trinidad and Tobago service was also flown. CPS Jetsave was offering a whole range of connections to Central and South America and the Caribbean.

Also during March 1974 the company clocked up another 'first' when the Civil Aviation Authority granted the first ever IT-affinity group charter licence for Gatwick to Hong Kong flights. Whilst the Comet, 1-11, 727 and 748 fleets were busy on their many and varied tasks, the 707s were also very active. They too took their share of Inclusive Tour work, but the operation of the trans-Atlantic ABC flights from Gatwick, Manchester and Prestwick was their major occupation. These services were further boosted in the autumn of 1975, when the airline was awarded a £4.7 million contract by International Weekend Inc. of Boston for the carriage of over 50,000 passengers from Boston to Gatwick and return for seven and fourteen-day holidays between March and November 1976. Again, up to seven flights a week were planned!

The Boeing 707s operated many of the ABC flights, along with many *ad hoc* charters to destinations as far afield as Los Angeles, New York, Singapore, Sydney and a number of desinations in Africa. By now the airline's Inclusive Tour charter network was the largest of any British airline, with departures from Aberdeen, Belfast, Birmingham, Bournemouth, Bristol, Cardiff, Castle Donington (East Midlands), Gatwick, Luton, Manchester, Newcastle, Tees-Side and West Berlin.

The summer 1974 timetable. (14 April - 13 Oct)

MODERNISATION

It was during one of the African flights that those aboard one 707 en route to Mauritius found themselves being escorted by a Soviet-built MiG jet of the Somali Air Force into Mogadishu for a supposed infringement of that country's air-space! The 83 passengers aboard were accommodated overnight and the next day the Captain (representing the Company) was fined £600; it seems that due to an administrative oversight, not all the correct papers had been filed with the appropriate Somali authorities five days in advance!

Fashions and styles change, so for the start of the 1975 summer season a new uniform for all stewardesses was announced. Designed by the House of Mansfield, it was a two piece worsted suit in navy blue with red trim. Although this new outfit was met with praise from all, it did have one drawback - wool was very hot in the warmer climes! But that jumps ahead of the story!

Captain Keith Moody recalls an incident on one of the long haul 707 flights that seems almost unbelievable, but it does demonstrate the persistence of the Royal Canadian Mounted Police...

"I had taken a 707 into Toronto where we had a two-day stop-over. As it was mid-winter and some of the cabin staff had never been to Niagara Falls (which are particularly beautiful with all the spray frozen into fantastic patterns) we decided as a crew to visit and advised the handling agents of our plans.

We were all out of uniform and had just called into a snack-bar to de-frost when I felt a hand descend upon my shoulder and a Canadian voice say *'Captain Moody*

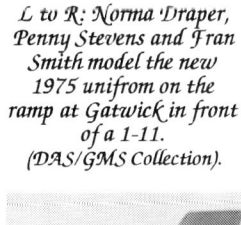

L to R: Norma Draper, Penny Stevens and Fran Smith model the new 1975 uniform on the ramp at Gatwick in front of a 1-11. (DAS/GMS Collection).

of Dan-Air?' I looked around and there was this huge 'Mountie' in full uniform!

It seems that another company 707 from Vancouver was about to stop at Toronto and they needed us back at the airport pronto, and the only way they could think of finding us was by calling the RCMP; up till then I never really believed that the 'Mountie always gets his man' - but that one did!"

Then there were the 'Haj' flights, carrying Mahommedan pilgrims from all over the world to join in the ceremonies at Mecca which every believer is supposed to perform at least once during his lifetime. These charters were very intensive and, in order to smooth out the process, the Company despatched Operations Manager Keith Balston out to the Middle East for the duration of the pilgrimage. Captain Moody also remembers some these flights, if only for the intensive flying and the number of passengers carried.

"One year we were operating a single 707 out of Teheran on behalf of Iran Air and we flew around the clock carrying 23,000 passengers to Mecca, returning empty to pick up another load after changing crews. This went on the the entire month of the pilgrim season. After it was over, the whole thing reversed to get the pilgrims back to their homes.

Many of the world's airlines were involved in the haj flights - they were good business. I remember one airline brought pilgrims from Indonesia! During this same season we also had a Comet flying 'haj' out of Kabul and later another season out of Nigeria.

The 707s were placed into greater and greater use when, during the 1970s, a series of 21 day round the-world flights calling at 19 countries was operated for a German tour company calling in, amongst other places, at Bankok, Tokyo and Honolulu. It was on these flights that the cabin crews individuality rose to the fore, with the girls wearing unofficial uniforms for certain sectors - Japanese

Five cabin staff get lei'd! Proof that the girls did wear unofficial uniforms on the 707 'round the worlds' comes with this snapshot from Norma Franks showing the stop-over at Honolulu. (Norma Franks)

MODERNISATION

style jackets out of Tokyo and Hawaian Mumu dresses into and out of Honolulu!

Keith Moody was the Captain on one of the 'round-the-world's' with 707 G-AZTG in 1976 that, although managed to maintain its schedule, was far from uneventful...

"It was a great privilege to be offered a world tour on the 707. The German passengers were demanding and expected punctuality coupled with good service. To achieve this required good crew co-operation for we would all be in close proximity for a period of three weeks!

The cabin staff (Pat Marsh, Linda Renddina, Val Jackson, Eve Strode and Petra Iwan) would be the mainstay of the operation, not only giving good service en route, but also developing a rapport with the passengers. Then there was the flight crew. First Officer, Tubby Greif was a good pilot with a lovely sense of humour. Flt. Engineer Lewton was a real professional who, off duty, would entertain us all by playing any instrument to hand and singing. The Navigator, Jamie Jamieson, whose accuracy across the Pacific impressed us all and finally, our ground Engineer Gordon Richards, whose skills would be tested to the full at Honolulu. A grand bunch to be with for 21 days, let alone to be paid for the pleasure!

What memories that trip held: the red forts of Agra, the Taj Mahal; the beautiful petite girls of Thailand and their floating markets; the magnificent temples of Burma and the excitement of bustling Hong Kong and Taipei. Tokyo - what a busy airport - aircraft parked wingtips almost touching - so busy, in fact, that we were requested to park the aircraft at Fukuoka, 550 miles south-west of Tokyo, close to Nagasaki!. Then that interesting sector across the Pacific, passing abeam Midway to Honolulu - a blissful climate and a ferry trip around Pearl Harbour, visualising the horrific Japanese air attack all those years ago.

After three restful days in Honolulu our take-off to Oakland in California was interrupted by a hefty turbine failure on No. 3 engine that required our return to the terminal. Operations quickly managed to charter a DC-8 that was due to position empty from Tokyo to Oakland to take our passengers and fly a replacement engine out to us (the nearest we had was in New York). We set off the following day and rejoined our passengers there. After a one hour turnaround it was off eastwards again, heading for Niagara where the passengers spent three hours enjoying the

The programme for Captain Keith Moody's 'round the world' flight in Boeing 707 G-AZTG. The times shown are in 'Zulu', that is Greenwich Mean Time (GMT). Therefore to get an accurate picture of arrival and departure times, timezone differences must be taken into account...

Fri. 27/2.	Sat. 28/2.	Sun. 1/3 to Tues 2/3	Wed. 3/3	Thurs 4/3 to Fri. 5/3	Sat. 6/3	Sun 7/3 to Mon 8/3	Tues 9/3	Wed 10/3 to Fri 12/3
Munich 13.00 Z to Teheran 20.15 Z	Teheran 14.05 Z to Delhi 20.20 Z	Delhi	Delhi 03.30 Z to Rangoon to Bankok 13.05 Z	Bankok	Bankok 06.30 Z to Kong Hong 10.15 Z	Hong Kong	Hong Kong 0400 Z to Taipai 05.30 Z to Tokyo to Fukuoka	Fukuoka

Sat 13/3	Sun 14/3 to Mon 14/3	Tues 16/3	Wed 17/3	Thurs 18/3	Fri 19/3	Sat 20/3	Sun 21/3
Fukuoka 09.20 Z to Tokyo to Honolulu 19.10 Z	Honolulu	Honolulu 23.00 Z	Oakland 03.55 Z	Oakland 14.30 Z to Niagara to New York 00.20 Z	New York	New York 23.15 Z	Munich to Gatwick 11.05 Z

...the programme shows the very minor delay to the aircraft (but not the passengers) caused by the blown engine at Honolulu. Dan-Air made two of these flights a year in the mid-1970's. Each could be regarded as an epic flight!

DAN-AIR SERVICES

```
LGWOODA
.HNLNVXH 170045
1/DA8008 GAZTG
2/ PHNL
3/MINIMUM 24HRS
4/NBR 3 ENGINE FAILURE
5/NBR 3 TURBINE DISINTEGRATED ON APPLYING TAKE-OFF POWER
6/ONE JT4A-12 POWER PLANT BUILT TO NO3 POSITION
AND ALL EQUIPMENT NECESSARY TO CHANGE ENGINE
7/NONE REQUIRED
8/NONE. GROUND ENGINEER RICHARDS HAS FULL COVERAGE
9/WAIKIKI SURF TELEPHONE 672-9235
10/RAMADA INN TELEPHONE 841-8031
11/CONSIDER POSSIBLE TRANSPORTATION PAX BY SCHEDULE CARRIER TO OAKLAND
12/REQUIRE T/C MINIMUM 4000 US DOLLARS TO BE SENT TO LOCAL BANK ASAP
GROUND ENGINEER CONFIRMS NO ENGINE AT THIS STATION
REGARDS
CAPT R J MOODY
```

The 'Aircraft On Ground' signal sent by Captain Moody to DANOPS Gatwick explaining the engine problem at Honolulu.

magnificent falls. Then airborne again to New York for our last nightstop together as a crew. The next day - a Saturday departure - direct to Munich, to arrive one hour ahead of schedule and then position back to Gatwick.

A very satisfying trip, with a marvellous crew. They are never-to-be-forgotten memories, even though seventeen years ago!".

But back to more mundane things! On 20 April 1974 a scheduled Newcastle - Gatwick service was opened, jointly promoted with British Caledonian Airways. This was initially flown twice daily by Dan-Air using Comet aircraft, but they were replaced with BAC 1-11s during November. A further Newcastle service was opened during April with a twice weekly Isle of Man flight. This route was flown on Saturdays and Sundays during the summer, but was restricted because the route was flown also by British Island Airways Heralds.

The dramatic increase in fuel prices resulted in the application of surcharges during 1974 and, coupled with the dramatic collapse of Clarksons Holidays Ltd and its associated holiday airline Court Line much dislocation had to be rapidly tackled by the staff. Luckily, the airline's policy of spreading its commitments over a number of different types of operations meant that Clarksons only constituted some 11% of the airline's business. Following the

BAC 1-11 G-ATPJ in an early 1-11 colour scheme taxies in. (DAS/GMS Collection).

MODERNISATION

Awaiting passengers at a rather snowy Aberdeen is 1-11 G-ATTP.
(DAS/GMS Collection)

collapse of Court Line, four of their BAC 1-11 500s were obtained, re-painted and re-registered for use on Dan-Air's IT flights previously been flown by Comet 4s and some 4bs.

After the traumatic events of 1974, tour operators took a more realistic view of the amount of capacity they placed on the market, which resulted in much better performance from Dan-Air, despite the continued weakness of sterling. Although unpopular, the fuel surcharges were accepted in the main by the travelling public as being reasonable under the circumstances.

This more sensible policy allowed Dan-Air and its associated engineering subsidiary to expand both in terms of size and profitability so that Mr Newman could state during the AGM held on 5th May 1977 that in 1976

> "...turnover increased by 50% as a result of several factors, including the extra capacity offered, additional aircraft engineering business and the decrease in the value of sterling, which in turn raised the price of fuel. Our scheduled services are expanding and our co-operation with several leading oil companies for charter services to the Shetland Islands linked with North Sea oil exploration have increased.
>
> During the winter months long-haul freight operations were carried out to all parts of the world with two Boeing 707-321C aircraft, which had been acquired during the latter part of the summer"

Dan-Air tried everything with the 707 in order to make it as profitable as possible. They were used for passenger charters during the peak of the summer season and then the passenger configuration interiors were stripped out so that the aircraft could be used on freight contracts during the less busy months. Sadly, one of the 707 freighters was not to survive for long, for through no fault of the crew or the Company, on 14 May 1977, G-BEBP was lost when the stabiliser failed as it approached Lusaka Airport. All six crew members aboard lost their lives.

DAN-AIR SERVICES

This accident (the cause of which had been totally unforeseen) was to have not only a tremendous effect on all within the airline, but offered profound implications for the industry as a whole. The aircraft concerned had been built in 1963, delivered to Pan American Airways as their first freighter, and had flown about 47,000 hours. The 707 stabilizer, like many other parts of a modern airliner's structure, was thought to be 'fail-safe' - that is, if a part of the structure did fail, it would do so in such a manner that the failure would not endanger the aircraft before discovery during the next scheduled inspection.

Investigations revealed that Dan-Air had maintained and inspected the aircraft in the correct manner, so no blame could be attached to them: the problem lay in the basic design. The heart of the problem was the definition of 'fail-safe' - did it mean 'having an indefinite life maintained by inspection and maintenance' or was there a point in the life of a structure where 'fail-safe' equalled the

Boeing 707 freighter G-BEZT with the additional titling on the fuselage roof showing that the aircraft was 'operated on behalf of IAS Cargo Airlines'. (M Forsyth Collection).

A beautiful air-to-air photograph of BAC 1-11 G-BDAS in one of the intermediate colour schemes. (DAS/GMS Collection).

MODERISATION

A pair of company BAC 1-11s at Aberdeen Airport. The closest aircraft, G-ATTP is in the older scheme, the aircraft behind is in the scheme that replaced it. (DAS/GMS Collection).

'Safe-Life' no matter what was done?

Dan-Air and Boeing quickly resolved the immediate problem with the remainder of the airline's 707 fleet, but the overall problem took much study and many years to resolve; this is the main reason why so many airlines attempt to keep their airliner fleets as young as possible.

The shortage of aircraft following the accident was made good at short notice by leasing an additional 707 freighter for a year and a second aircraft was purchased in the autumn as a permanent replacement. Inevitably, these arrangements proved more costly, but enabled the long-haul freighting operation to continue with the minimum loss of time.

1977 was another year when the peak summer air travelling period was badly disrupted by air traffic control delays. In August a series of 'go slows' by the air traffic control assistants soon developed into a full strike and this, coupled with sporadic disruption from air traffic controllers in France and Spain, continued for ten weeks. This led to delays of up to 24 hours, causing much hardship to the hard-pressed travellers who just wanted to get away on holiday.

DAN-AIR SERVICES

A Rolls-Royce Spey recieves attention at the Manchester maintenance facility, whilst in the background can be seen the engine nacelle of 1-11 XR. The 'hush kit' silencer pipe can be clearly seen at the rear. (DAS/GMS Collection).

Modifications and conversions.

By the late 1970s many airlines were beginning to respond to growing complaints about aircraft noise from those living near airports. Efforts in 'good neighbourliness' resulted in studies to quieten some of the older jet designs, but these studies appeared to demand that noisy aircraft would have to be scrapped, or at least re-engined with more modern, quieter engines. In an effort to extend the life of the 1-11, the British Aircraft Corporation developed a 'hush-kit' for application to the notoriously noisy Rolls Royce Spey, comprising intake and by-pass duct lining, an extended, acoustically-lined jet-pipe and a six-chute jet mixing silencer. Dan-Air's Manchester engineering base worked closely with BAC and started a modification programme to fit these 'hush kits' not only to the airline's own fleet, but also to many other airlines 1-11s.

One unusual conversion undertaken at Manchester was the modification to a Greyhound Leasing 1-11 on loan to an Arab sheikh who wanted the interior to match a suite at the Brussels Hilton - complete with luxury bathroom, bedroom and lounge!

So, as the end of the 1970s approached, the airline was continuing to expand, surviving the problems of the oil crisis and many of the troubles that dogged the Affinity Group Charter market. A new decade was dawning, and with it, new challenges.

CHAPTER SIX
Pax, Cargo, Oil and the 748

The discovery of major oil and gas deposits off the Shetland Isles in the early 1970s and their subsequent development led to a dramatic increase in air traffic to Sumburgh, then the islands' only airport. The number of passengers using the airport rose from less than 95,000 in 1973 to over 685,000 just five years later, much of which was generated by Dan-Air.

Early in 1974 one of Dan-Air's 748 fleet was based at Aberdeen's Dyce Airport for use on oil-rig charter work, with a contract held with the Conoco Oil Company to ferry workers between Aberdeen and Sumburgh. Further contracts were obtained from the oil companies so more 748s were moved to the airport. When not involved in direct oil supply work, they were used on charters from Aberdeen and Edinburgh to many Scandinavian and European countries; Amsterdam, Kristiansand, Oslo, Bergen, Esbjerg, and Rotterdam being just some destinations.

But this was not the first time that Dan-Air had become involved in oil-rig work: during 1969 at least one of the Ambassador fleet was used for a number of oil and natural gas support charters to Aberdeen from Norwich (Horsham St. Faith) airport.

Aberdeen was now rapidly developing as one of Europe's major centres for both oil and gas exploration, so the Company's 748s roamed much further afield than Great Britain - a contract for a regular weekly charter flight between Aberdeen and Oporto in Portugal was operated during 1974 and 1975 carrying an exchange crew for the Transworld 61 rig.

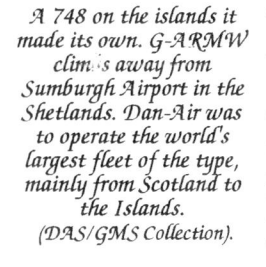

A 748 on the islands it made its own. G-ARMW climbs away from Sumburgh Airport in the Shetlands. Dan-Air was to operate the world's largest fleet of the type, mainly from Scotland to the Islands.
(DAS/GMS Collection).

DAN-AIR SERVICES

The Aberdeen outstation, based on the east side of the airfield in a wooden building that used to be the CAA Fire Station peaked in workload from 1978 to around 1980. At its height 36 crews operated fourteen 748s with three Operations Controllers under superintendent Bob Willis. Initially the aircraft were flown with flight deck crew plus a loadmaster to look after the cargo. When the scheduled service to Gatwick was inaugurated, cabin staff were brought into the picture and gradually drifted across into Oil Support work.

From about 1976 the outstation operated virtually independent of main base; it had its own engineering support team of about 20 members who, out of necessity, had to find unusual solutions to the working conditions. Aberdeen was very busy and Dan-Air could not acquire hangar space; in order to complete an engine change in relative comfort, the engineers designed and built their own Engine Change Shelter, equipped with heat, light and power, that could be wheeled up to and over an engine of a 748 whilst parked on the ramp to allow work to proceed!

The Company's oil-related support flights in Scotland rapidly evolved into an enormous operation which brought with it the need to obtain more aircraft to cope with the traffic. Since its conception, Dan-Air had always followed a policy of avoiding the enormous expense of obtaining brand-new aircraft if suitable second-hand machines in good condition were available at the right price. So it was with the 748. Peter Carter was tasked with finding examples that could be obtained at the right price. In Argentina, a number of machines were available, so Peter flew out, inspected them, and arranged the purchase and overhaul so that they could be flown back via the Caribbean, North America, Canada and then 'over the top' to the UK.

G-ARAY in the final colour scheme flies over a rather bleak Scottish landscape that has recieved its first dusting of snow - a typical scene for the start of winter operations 'Up North'. (DAS/GMS Collection)

PAX, CARGO, OIL AND THE 748

Captain John Smith at the controls of G-BEKE. John assisted in bringing this and a number of other 748s back from Argentina. As these aircraft were not equipped with auto-pilots, they had to be 'hand-flown' all the way!

Dan-Air was used to obtaining aircraft from all over the world, but these Argentinian 748 ferry flights must be regarded as epic journeys. Captain John Smith, who had been flying the 748 since the very early days, assisted on the ferry flights back to the UK:

"I flew G-BEKF on 26 March 1977 from Commodoro and 'KE from Cutralco into Buenos Aires on the 29th for engineering work to be done before they could be brought back. We (myself and First Officer Hanton) left Buenos Aires on 5 April in 'EKF, routing via San Paulo, Brasilia, Belem, Paramacibo, Barbados, San Juan, Palm Beach, Wilmington, Bedford, Goose Bay, Frobisher, Sundestrom, Kulusok (where we had to divert to fix a tech problem), Reykavik and finally Manchester, arriving there on 14 April! The total ferry time from Buenos Aires to Manchester was fifty-six and a half hours, all of which were 'hand flown' as these Argentinian 748s had no auto-pilots!"

Other examples were bought or leased from all around the world, including one, operated as G-AYYG, from New Zealand! Known affectionately by the crews as 'Lil' (from the Mount Cook Airlines Lily emblem on the tail) the aircraft made three round-trips from New Zealand to Scotland, spending the summer months in northern climes before flying south for the winter!

Eventually fourteen 748s were based in Scotland, serving the oil industry in many ways, but primarily as the fixed wing link between

G-AYYG 'Lil' - a Mount Cook Airlines 748 leased for three summer seasons seen parked on the ramp at Aberdeen in 1981. The only change in colour scheme was the application of the 'Dan-Air London' titles on the fuselage - the stripe and tail were still painted in 'Mount Cook Blue'!

DAN-AIR SERVICES

Aberdeen, Glasgow and the Shetlands, where oil-rig workers transferred to helicopters for onward flights to rigs in the oil-fields. The scale of the operation was enormous. Traffic (and number of passengers carried) steadily rose, so that by 1980, when the peak had just about been reached, Dan-Air made about 6,700 flights from Aberdeen and Glasgow to the Shetlands.

Captain Smith (known to many as J.G.) remembers very clearly what it was like operating out of Aberdeen and some of the lengths they had to go to to fly...

"There always seemed to be a pioneering spirit; everybody mucked in to get the job done. There where times when we came up with some strange answers to problems though. We were very busy on the oil work, and had to get as many flights a day in as possible. Sumburgh opened at 0800 in the morning and we planned our first flight in to arrive as soon after that as possible. We needed to know the actual weather conditions there before we left Aberdeen - but there was no 'Met' available until the airfield opened! Our solution was to ring another pilot, Alan Whitfield of Loganair, who lived in a cottage near Sumburgh. On receiving our telephone call he would peer out of his bedroom window and give us the Sumburgh 'actual'!

This worked fine until we got really busy, when we started so early, and he got so many calls he was not getting any sleep. Evenutally we found that the Sumbugh Lighthouse keeper was prepared to help, so we thoroughly briefed him on what we wanted and provided him with a simple set of rules. If he could see the airfield, which was around 3,000 metres from his position, we were O.K. to go. If he could only see the Sumbugh Hotel (1,500 metres away) it was decidedly marginal, and it was best that we waited awhile. If he was in cloud, then forget it, time to go for another breakfast!"

A dramatic photograph of a 748 coming into land on Runway 33 at Sumburgh. At the peak of the oil flights there was a procession of Company aircraft arriving and departing here.

PAX, CARGO, OIL AND THE 748

Scottish staff and their aircraft. Left to right: Loadmaster Norman Port, F/O Dick Whittaker, Engineer Jack Inkster, Capt Tom Buckland and Traffic Officer Liz McCullough.

Of telephones and passengers...

Aircraft operations in the far north of the United Kingdom could be, and were, very different than in the more formalised south. All those involved had to think on their feet and quickly solve whatever problems life threw at them, as Airport Duty Manager John Fraser recalls:

"Two incidents spring to mind as to just how different it could be. I remember well the annual Sullom Voe 'Airlift' that started just before Christmas. I was a Traffic Officer at the time and know that for one of the airlifts in the late 1970s we had something like 6,500 construction workers carried on 130 charter flights to get off the islands and back home to their families in time for Christmas - and after the New Year we had to get 'em all back again!

That number of people going through the old passenger terminal in a few days would have literally swamped the airport at Sumburgh. An apparently desperate situation called for a desperate solution, so we hired a village hall a few miles away to use as our own 'check-in terminal' and then ferried the passengers by coach straight from there to the aircraft waiting on the ramp.

Unfortunately the hall did not have a telephone from which we could ring the numbers and weights through to those waiting with the aircraft at Sumburgh, but the GPO said there would be no problem in laying on a temporary phone (this was in the days before personal 'cell-phones'). We thought all had been arranged, but two days before the Airlift started suddenly there was a problem - it appeared that the main telephone cable was on the other side of the road from the hall, but the GPO said that they would do their best.

Come the big day and hundreds of construction workers started arriving at our newest 'Check-In' Hall to be processed through and then board coaches for the airport. Then came the interesting bit - the moment each coach left, one of us had to cross the road, carefully climb a barbed-wire fence into a field full of sheep, lift a plastic bag that covered a normal domestic telephone and ring the airport!

DAN-AIR SERVICES

These were certainly times when pressures of the job to get aircraft away forced the staff into esoteric forms of 'lateral thinking'. During another incident, John must have remembered back to the Ambassador episode at Rheims a few years earlier:

"On another occasion I had to get the waiting-to-depart passengers to give a 748 a push-back! This somewhat strange event happened in the days when aircraft traffic at Sumburgh was at its height and parking space was very much at a premium before the new terminal was built. Things were busy (as usual) and we had about half a dozen 748s parked nose-to-tail taxi-cab fashion on the ramp awaiting incoming helicopters from the oil fields, near to the temporary accommodation we were using. Suddenly we noticed that a light aircraft pilot had parked and locked up in front of our next-away 748 and that he had parked too close for the 748 to turn to clear his aircraft.

Our contracts with the oil companies called for strict departure times - and even if they hadn't, the passengers wanted to get home as quickly as possible after their duty period offshore.

The only solution was to go over to the other side of the airport to locate an aircraft tug to 'push-back' our aircraft. Getting a tug from the other side of the field could take quite a while as it had to cross the active runway and just as I was just discussing it with the skipper, some of the passengers overheard our conversation. Being experienced passengers they knew that the third helicopter had arrived and they wanted to get away quickly so they volunteered assistance. I knew a 748 could be man-handled on the ground, so after further consultation with the aircraft's skipper, I got a number of burly oil-workers to do a human-powered 'push-back' of about fifteen feet - or at least enough to turn the aircraft!"

Into Scatsta...

As the East Shetland Basin pipeline terminal and tanker loading complex at Sullom Voe was nearing completion, a requirement arose for a closer air-head than Sumburgh, which had by now almost reached saturation point and anyway, was some two hours away by road.

Back in 1940 an airfield had been literally hacked out of the peat

Into Scatsta! The aircraft is on finals and has just passed the Sullom Voe Terminal complex. To the right of the runway are the Terminal buildings built, along with the parking ramp on the remains of the old wartime runway. At least six 748's could be found on the ground here during the peak times.

PAX, CARGO, OIL AND THE 748

at nearby Scatsta as a land airfield to support the major flying boat base in Sullom Voe, although the airfield fell into disuse soon after the war. 38 years later, on 22 July 1978, after the Local Authority approved the rebuilding of one of the wartime runways and the construction of limited terminal facilities, Dan-Air's Captain John Smith brought HS 748 G-ARRW down onto the sloping 06/24 runway and into the refurbished airfield for a proving flight. Two days later Capt. Smith made the first-ever commercial flight into the airfield and services have continued ever since, although nowadays at a much reduced rate. As John recalls;

"We found the 748 ideal for the Scatsta run. Normally we used 44 seaters, but some aircraft were equipped with 21 first class seats in the rear of the cabin and a further 12 economy in front of a dividing bulkhead. There were no payload limitations out of Aberdeen and within reason, this also applied to Scatsta, despite its short runway length. At the height of the operation into this airfield we had three machines based at Scatsta and one in Glasgow, the latter airport handling most of the traffic from Sullom Voe. After the oil terminal was completed the flights were greatly reduced and we then based the BP contract aircraft in Aberdeen.

July 1978 saw the oil flights reach a peak that was sustained for many years and can be used as a good example of the intensity of flying that was conducted during a typical month in the 'far north'. On the 3rd myself and First Officer Clift took G-BEKG from Aberdeen to Sumburgh twice, doing the same run again on the 6th with 'KF in the morning and 'JE in the afternoon, accompanied by F/O Backhouse. On the 7th myself and F/O Silver flew twice up to Sumburgh from Aberdeen in 'KG for route and line checks.

8 July saw myself and F/O Yates in 'YG fly Aberdeen to the Isle of Man and then on to Bournemouth. The next day we took G-ARAY from Gatwick to the Isle of Man and onto Aberdeen, then it was back on the Aberdeen - Sumburgh run; one on the 10th and 11th with 'KD and F/O Asher and 'AY and F/O Hertszburg respectively. I had a couple of days off, then made three more trips on the route on the 14th using 'JD with F/O's Asher and Kilkelly. Halfway through the month now, and already visited Sumburgh eleven times; the 15th was no different - with 'YG. The 17th was different, a 1.55 flight around Aberdeen, checking out 'JE for Certificate of Airworthiness renewal. The 19th and 20th saw four more flights to Sumburgh in 'YG but then it was something very special. On 22 July I positioned G-ARRW along with F/O Barnett from Sumburgh into Scatsta for the first time, ready for training. Scatsta is the smallest airfield in the UK that can accept a 748 and all crews have to make at least one trip before flying in 'for real'.

We made a couple of Scatsta - Sumburgh flights on the 22nd before making the

Spot the trees! G-BEKE parked on the ramp at Scatsta. The photograph shows well the bleak landscape - it could be very windy here at times!

DAN-AIR SERVICES

first and second commercial flights from the airfield to Glasgow and back on the 24th. The rest of the month, the 26, 27th and 28th was spent on Scatsta-Glasgow, but on the 2nd we had to turn back to Glasgow on the return due to bad weather. Luckily, that never happened that often!

In all, I was airborne for 81 hours 40 minutes during the month, using seven different aircraft on 35 different trips, only one which we failed to complete - and that was due to the weather!"

Flying into the Shetlands was not only 'interesting' for the pilots - it could be equally challenging for the stewardesses and passengers alike! Mo Perera alternated her flying on Boeing 727 charters to the sun with looking after the large number of oil industry personnel Dan-Air was carrying up to Sullom Voe and back.

"The Shetland weather could make landing at both Sumburgh and Scatsta difficult at times, especially in the winter, and on the 44-seat 748s we only had one stewardess to look after the passengers. Very often even the big, apparently tough oil business passengers would be glued to their seats as we approached the Shetlands in a gale, showing lots of white knuckles. It seemed as if it was always the big guys who got the most terrified!

Occasionally the aircraft would overshoot the runway and go around again, and sometimes would even make three attempts to land and still have to go back to Glasgow to try again the next day".

Despite the very challenging flying environment and the ever-present possibility of getting stuck in Scatsta due to inclement weather it did have its compensations. Mo again:

"We had one special Saturday charter from Scatsta to Edinburgh and back, carrying senior personnel to watch an international rugby match at Murrayfield. A lot of extra spirits and drinks were ordered, and when our passengers arrived back in Scatsta after the match, much was left over. That night a terrific gale sprang up and, for want of a better place to stow them, I had the dozen bar boxes (along with the spirits!) taken to my room in the local hotel for the night. On the Sunday, the 748 was blown from its moorings at the aerodrome and damaged, but the charterer gave me instructions to hold a party on Monday for the crew and their personnel so as to finish it all off - and we did!".

The main cabin of 'KE coming back from Scatsta.. 'Oil Related No.1' Cabin Staff Pamela Morrison tends to the passengers.

PAX, CARGO, OIL AND THE 748

'Link City' developments

On 4 April 1972 HS748 G-ARAY operated a proving flight over the new Luton - Leeds - Glasgow service, which was not officially inaugurated until a week later on 11 April. On 5 April G-ASPL opened a Bournemouth - Birmingham - Liverpool and Manchester - Newcastle service which was followed on 27 May with a once-weekly Swansea - Jersey 748 service, the first time that Swansea had been served for many years. On the same day a Newcastle - Carlisle - Jersey route was opened and further Channel Islands routes opened from 1 July when Dan-Air took over the former Channel Airways Bournemouth to Guernsey and Jersey services.

Most of the Isle of Man schedules for the 1972 season were flown under sub-contract by a Vickers Viscount from Kestrel International. Thus on 27 May, Viscount G-AVJB re-opened the Newcastle-Carlisle-Isle of Man route. In March 1973 the Company added Tees-Side to its Link City network from Bournemouth, Bristol, Cardiff and Newcastle. A further service between Tees-Side and Amsterdam was added on 2 April and in May a 748 Ashford (Lympne) to Jersey weekend service was flown.

The scheduled and charter services between Gatwick and Ostend had been continuing for a number of years, but from 20 May 1974 the service was operated from a pool in conjunction with the Belgian airline Sabena. 44-seat 748s were used twice daily in 1974, but this was reduced to daily in 1975 under a joint Dan-Air/Sabena flight number. 1974 also saw the 748s replaced on the Gatwick - Clermont Ferrand and Montpellier service by Comets and later 1-11s. The Cardiff - Bristol - Amsterdam service was re-opened, after lying dormant for many years, with a series of 748 flights.

For some time since the take-over of Skyways, Dan-Air had been concerned about the future of operations from Ashford (Lympne) as the airfield had runway length limitations and an indifferent weather record. This was a good enough reason to leave,

On the 'Link City' route! 748 G-ATMJ during turn-around at Newcastle Airport. This colour scheme was regarded as the third on the 748 fleet.. (DAS/GMS Collection)

DAN-AIR SERVICES

On the ramp at Gatwick! Here G-AXVG is seen basking in the late afternoon sun. This aircraft was used on both the 'Link City' and overseas 748 services. (DAS/GMS Collection)

so on 31 October 1974 Dan-Air closed down the airport and moved all its cross-channel operations to nearby Lydd, with all 748 maintenance done at either Lasham or Manchester.

John Varrier joined Dan-Air in April 1972 when the airline took over his employers, Skyways International:

"For a year or so I was Personnel Manager at Ashford Airport, and then took over as Manager and Budget Controller. My responsibilities were then increased to include the French operation, and I commuted a couple of days each week to the Paris office. I then oversaw the move from Ashford to Lydd, and shortly afterwards was moved again, this time to the London Office in the Scheduled Services Division. I saw the Scheduled operation grow from a handful of HS748s to a major commercial activity, using a growing part of the jet fleet".

Coach-Air services.

The 'Coach-Air' concept had stood the test of time, for as long ago as 1955 the first service was flown out of Lympne, using a Skyways Dakota. HS748s had been employed for a number of years but network increases placed demands on Dan-Air's 748 fleet, so additional capacity had to be found.

The airline had experimented with 1-11s on the Lydd - Beauvais schedule, but as they were found to be unsatisfactory, the idea was dropped. The 1974 winter Lydd - Beauvais operation was conducted with 748s, but from March 1975 onwards additional capacity was provided by making use of a leased ABC Viscount 800, supplemented during the peak summer months with a further

Showing both aspects of the Coach-Air services is Viscount G-ARIR and the East Kent/National 'Dan-Air 909' service. The small notice, just visible to the right of the main cabin door on the Viscount on the original photograph, states that the aircraft was 'leased from Alidair'. (DAS/GMS Collection)

PAX, CARGO, OIL AND THE 748

748 on hire during the weekends from the Civil Aviation Authority and flown with a CAA crew but with Dan-Air cabin staff. Luckily the CAA livery was broadly similar to that of the airline, which meant that apart from the removal of the fuselage titles, the only change was a small notice adjacent to the passenger cabin door declaring that the aircraft was on hire to Dan-Air Services! The ABC Viscount remained in use until the late autumn, when it was replaced with an earlier 700 series machine from Alidair.

During 1975 Liverpool was dropped from the Link-City network. This meant that the Liverpool - Amsterdam service closed after fifteen years; the service had been lately flown by Comets to try to pick up the load factors, but without success.

The summer of 1975 saw the introduction of the 1-11 onto the Gatwick - Jersey service on 18 May and the Newcastle - Bergen route two days later. The Newcastle - Kristiansand route became five times a week from 1 June and the Swansea - Jersey service was flown thrice weekly. An Aberdeen to Isle of Man service was inaugurated by one of the Aberdeen based 748s on 24 May whilst a Gatwick to the Isle of Man route was opened later the same day. The Berne service from Gatwick became daily from 13 May and proved such a success that passenger levels increased by 80% over the summer months! The service to Clermont Ferrand was now extended to Perpignan. 1975 also saw the introduction of a nightly freight service to Belfast.

The 748 design proved very suitable for one unique piece of work undertaken by the Manchester engineering base that harked back to the very early days of cargo-charters as done by the DC-3s. The work was done to one of Dan-Air's own 748s specially

John Fraser overseas the unloading of a turbine engine out of the specially modifed hold of 748 G-BIUV through the oversized rear door.
(DAS/GMS Collection)

103

purchased for the job and involved the installation of an eight foot by eight foot cargo door in the rear fuselage, together with ball mats in the door entrance and roller mats along the entire length of the cabin in order to allow swift easy access of bulky or awkward cargo. This meant that the aircraft could carry 'standard' cargo, such as Land-Rovers, livestock or any bulky items on pallets, and also easily accommodate pipes and oil drilling equipment of six inches diameter and forty feet long, especially important when one considered Dan-Air's Scottish oil support programme.

The aircraft, G-BIUV was soon placed into service. An aero engine had to be ferried to one of the smaller Greek islands, making good use of the aircraft's short field capability. Since the aircraft was easily convertible back into a passenger configuration, it also took a party of journalists to Sweden to view the new SAAB SF 340 airliner. On the journey back Dan-Air's public relations department made sure that there was an ample supply of brochures on board the newly convertible 748!

Of Cargos and Charters...
Meanwhile, not everything that travelled by Dan-Air 'Link City' network had two legs and strapped itself into a comfortable airliner seat. Lobsters, salmon and mushrooms were consigned to French and Belgian restaurants from Aberdeen; Cardiff sent tropical fish to Belfast and lobsters along with live eels went from Bristol to Amsterdam. Also from Bristol to Glasgow and the Orkneys went shipments of whey and cheese starters for local dairies. The Cargo Division arranged for the carriage of just about everything!

With ever-changing markets, any airline is sensitive to over-capacity and, with the early indications of a decline in oil-related charters, new uses had to be found for the 748 fleet. A number of machines were placed for use by overseas airlines. During 1981

The Night Mail. The interior of a 748 fully loaded with mailbags netted and roped securely. (DAS/GMS Collection)

PAX, CARGO, OIL AND THE 748

Passengers board a 748 at Cardiff-Wales airport in 1982. (DAS/GMS Collection).

British Airways announced a change in policy and aircraft used on their Scottish routes, so three Dan-Air 748s were leased to BA on highly satisfactory terms. These leases were renewed for 1982.

Dan-Air also became an integral part of the Post Office's drive to speed up the delivery of letters during 1979, called *'Spokes to Speke'*. Liverpool's Speke Airport was chosen by the Post Office as the hub for their nationwide distribution system because of its central geographic location. Every evening aircraft from the Dan-Air 748 fleet brought sacks of sorted mail into Speke from all parts of the country to have other sacks re-loaded from other parts of the country for forwarding on to their final destination.

Charters were not just the carriage of plane-loads of holiday-makers, as one charter during 1983 demonstrates. Crown Machines Ltd of Birmingham were specialists in the sale of high-technology computerised machine tools and flexible manufacturing systems, much of which was manufactured by Mori-Seiki of Japan. A major part of Crown's business was selling these machine tools around Europe so, in order to offer better back-up and spares service to their European customers Mori-Seiki established a spares and technical centre in Dusseldorf in 1982.

The announcement of the opening of this complex was of significant importance to Crown Machines and existing Mori-Seiki customers and it was realised that both Crown personnel, existing users and potential users should have the opportunity of visiting the complex at the earliest opportunity. Original plans indicated about 100 people wishing to go from Great Britain.

DAN-AIR SERVICES

After enquiries by Crown to local travel agents, it was evident that larger airline companies were unwilling to alter existing scheduled services to cater for this particular need. Crown wanted to make the trip in a day and although there were scheduled services that did allow this, it would have meant a visit time of only 30 minutes at the centre, which was obviously out of the question.

It was at this stage that Dan-Air's 748 Commercial Department became aware of the company's requests and quickly quoted times, prices and types of aircraft available. The anticipated number of passengers grew from 100 to 150, then to 200 and finally 230. Not only did the numbers increase, but some wanted to depart from different airports on different days to coincide with scheduled domestic flights from Glasgow, Dublin and Belfast.

Eventually Crown chartered five 46-seat 748 aircraft from Dan-Air on three separate days to make the journey to Dusseldorf and back from various airports around the U.K. As R.N. Pollard, the Managing Director, later wrote to the company in appreciation:

"The outcome of the visit was certainly a success and could be closely attributed to the efficiency and flexibility of the staff and crews of Dan-Air Services".

Other charters could be equally interesting. In July 1984 a 748 was hired to transport the wedding party of one Mr R. Taylor from Gatwick to Naples. The groom was Roger Taylor, drummer with the pop group 'Duran Duran', then at the height of their fame.

Understandably, the group wished to keep the arrangements secret from their fans, so special arrangements were made to whisk the VIP party from a private suite at Gatwick by coach to their waiting aircraft, where they were met with champagne and canapés. At Naples, similar arrangements were adopted and the transfer went unnoticed by the fans. The aircraft and crew remained at Naples for two days whilst the wedding celebrations took place, and were rewarded for their discretion with a block invitation to the reception, a nine-hour cruise 'around the Bay' by luxury cruiser. Just as discreetly, Dan-Air flew the party back to Gatwick!

Not all enquiries for charters were as successful. John Fraser recalls an incident where he could have kicked himself for missing the opportunity...

"When I worked in our Loading Office we dealt very closely with the 748 Commercial Department, particularly on freight flights - what we could carry, how much, how far - that sort of thing. One year, just before Christmas, Graeme Smith phoned me up and said they had received an enquiry about carrying reindeer to Lapland. I looked at the 'phone aghast... they had to be joking! He insisted it was true, but I was determined not to be taken in, because Graeme was a known joker. I told him that it was not a good idea, after all, their antlers would hit the ceiling, their hooves would go through the floor, we'd have to get pens built - no way. I put it down to the Christmas spirit and promptly forgot about it. Some months later I read in the paper about the film Santa Claus, starring Dudley Moore and they had shipped their reindeer back to Lapland after filming. I suppose we missed out on some good publicity there!

PAX, CARGO, OIL AND THE 748

A pair of Bandeirantes in the joint Dan-Air and Centreline colour scheme with the 748 'Charlie Juliet' behind pose for publicity photographs. (DAS/GMS Collection).

Joint arrangements.
The 'Link City' concept was further expanded when in 1982 Metropolitan Airways, a subsidiary of Alderney Air Ferries (Holdings) Ltd, began operating under the 'Link City' banner, with a pair of de Havilland Canada Twin Otters, G-BHFD and G-BELS. These aircraft were flown in full Dan-Air colours with the name *'Dan Air Link City'* over the entry door and the *'Metropolitan'* titles under the cabin windows and were operated on services between Bournemouth, Cardiff, Bristol, Birmingham, Manchester, Leeds/Bradford and Glasgow.

The agreement between the two Companies meant that Dan-Air provided reservation and other facilities for two years, after which Metropolitan would completely take over a number of routes. These plans were short-lived, for Metropolitan Airways suspended services on 31 August 1985 and went into liquidation on 2 September. A similar arrangement was entered into with Centreline for a short period in October 1982, using a pair of Bandeirantes (OY-ASL and G-CTLN) and a Cessna 310Q (G-AZVY) on services between Bristol, Cardiff and Glasgow.

Sadly, during the many years of Dan-Air 748 operation, two machines were involved in incidents that caused a tragic loss of life. The first occurred at 1601 hrs on 31 July 1979, when G-BEKF suffered problems on take-off from Sumburgh's 09/27 runway in the Shetlands and slid into an area of water known as the Pool of Virkie. Of the 44 passengers and 3 crew aboard, 15 passengers and 2 crew (both pilots) lost their lives, but 29 passengers and Liz Elphinstone, the air stewardess, survived. She was later decorated with the Order of the British Empire for her bravery in swimming to rescue many of the escaping passengers.

Two years later, at 1811 hrs GMT on 26 June 1981, G-ASPL was lost at Nailstone, Leicestershire, whilst engaged on a mail flight from Gatwick to East Midlands Airport, with the loss of both pilots and the loadmaster.

DAN-AIR SERVICES

748 G-BIUV at Gatwick on 23 March during the 1992 General Election campaign. As can be seen, this aircraft was chartered by the Liberal Democrats and carries their 'My Vote' slogan on the tail in place of the normal logo.
(Alan Edwards)

Twilight of the 748...
By the early 1990s Dan-Air's 748 fleet was firmly in the decline. The helicopters now used for oil support work could operate at increased range, which meant that the 748s were no longer needed and many were sold off. By March 1992 the night mail flights ceased, and the two aircraft used were stored at Manchester awaiting disposal. Both these machines were in need of a major overhaul that was deemed uneconomic and so awaited a potential purchaser. 'UV, the cargo convertible 748, was retained for any work that came along leaving G-BEKE for use on the Aberdeen to Scatsta and return long-term British Petroleum charter.

'UV achieved much public exposure during the 1992 Election campaign when it was chartered by Paddy Ashdown, leader of the Liberal Democrats, as his own personal election transport, and the aircraft was to been seen almost nightly in the television news broadcasts.

The 748 had proved to be a very useful aircraft for the airline to have on its books and, like the DC-3 many years before it, would be difficult to replace, but nevertheless the time had come to relinquish the 748 fleet.

By July 1992 the last six machines had been sold to Janes Aviation of Blackpool, although Kilo Echo was promptly leased back to complete the Aberdeen - Scatsta contract.

An end to the 748 era should have occurred at 1715hrs on 29 September when G-BEKE, operating as Flight DA7276, lifted off from Scatsta and set course for Aberdeen as Dan-Air's last 748 flight, but it was not to be. The infamous Shetlands weather played its hand for one last time, delaying the flight by 24 hours!

DAN-AIR SERVICES

CHAPTER SEVEN
Flying equals

Wherever a pilot learns to fly and whatever his experience, he must obtain a Commercial Pilots Licence and Instrument Rating in order to fly an airliner. A pilot qualifies on each aircraft type individually and renewal of the licences is required annually, involving a strict medical and a requirement for a minimum of flying hours logged during the previous year. Pilots start with any airline flying 'in the right-hand seat' with the rank of First Officer. Eventually, as experience, qualifications, hours and seniority mount the pilot progresses to the rank of Captain and the coveted 'left hand seat'.

When promoted, a pilot enters a whole new world of responsibilities. Like a Captain of a ship, his word aboard his own aircraft is law - though like a ship's captain nowadays he is not licenced to marry any runaways! The Captain is responsible for the safety of his aircraft and for the work of the flying and cabin crews. Departmental reports within the aircraft, be they concerned with flying, passengers or freight, metaphorically end in the lap occupying the left hand seat - the buck truly stops here!

All Dan-Air's flight deck crews have worked under the Chief Pilot, of which there have been remarkably few. The first was Captain F. R. Garside, then Captain John Cameron, followed by Captain Warren Wilson, Captain Bob Atkins, Captain Dick Spurrell, Captain Malcom Grant and finally Captain Mike Rothwell.

The Captain starts work before the aircraft takes off with the routine of pre-flight planning and cockpit checks. Work continues with the responsibility for the passage of the aircraft to its destination and for any changes in flight planning to be made during the flight. Even when on the ground at the desination, the work does not stop. Theirs is the overall responsibility for a proper turn-around or, if necessary, its hand-over to the Company Agent or another crew. The Captain's work only ends when the responsibility for the passengers, aircraft and crew has ended.

One of Dan-Air's Chief Pilots - Captain Bob Atkins. (DAS/GMS Collection).

During that period of duty, only a relatively small amount of time will be spent on what the pilot probably loves most - flying. For example, the Captain is responsible for the weight distribution of the baggage in the holds. He is also responsible for deciding exactly how much fuel to uplift and, within the overriding demands of safety, for practicing strict economy in this operation. This is also related to the weight of baggage, passengers and crew; tables, charts and computers

DAN-AIR SERVICES

greatly aid this process of precise calculation, but it is interesting to note that passenger weights are 'averaged'. On a Boeing 737-400 all males weighs 78kg, females 65kg, children 42kg and infants 10kg. There are also standard weights per item of luggage; on a domestic schedule each bag weighs 10kg, on a European schedule 12kg and on a charter service 13kg - over an aircraft load these work out remarkably accurately.

It was an accepted fact of life for every pilot flying with Dan-Air that they had to undergo a constant pattern of assessment of their professional abilities, by both outside agencies represented by the

A Boeing 737-400 load form that had to be completed before every flight

FLYING EQUALS

Civil Aviation Authority and internally by the Company's own training and assessment staff. There were also very regular medical check-ups and assessment of general health.

Recruitment and Flight Deck Management Training.
The airline may have found itself with a whole series of wonderful aeroplanes with all their splendid systems and a number of potentially highly skilled and highly profesional flight deck crews to use them, but nevertheless, each human member came complete with a whole series of very different, very human attitudes and foibles. How did the airline make sure that they operated the machines safely and efficiently?

The answer in one direction lay in very careful selection at the initial recruitment phase. Potential flight deck crews underwent a whole battery of tests, including Psychometric testing, which Dan-Air pioneered. Allied to this was Flight Deck Management training (FDM) the seeds of which were planted within Dan-Air in the days of the Comet. Members of management investigated a whole series of different systems that had been designed around the world, but none appeared to meet the airline's requirements fully.

Eventually Dan-Air developed its own series of courses under the overall title of Human Factors in Aviation which were copyrighted and sold to other airlines. From this foundation the Civil Aviation Authority developed its own courses which became mandatory for all airlines.

This selection and training ensured that no matter what the differences caused by human nature, the aircraft were operated to the highest levels of safety and efficiency. On a day-to-day level this was greatly assisted by the very early introduction of Standard Operating Procedures (SOPs). Dan-Air put in a tremendous amount of work in developing an entire manual that told staff how any particular situation should be handled. This really came about through operating a number of different aircraft types with crews that might have met each other for the first time on the day of the flight. With the introduction of SOPs, everyone knew exactly what their duties were in relation to each other in any given situation.

Females on the Flight Deck...
Over the years Dan-Air employed many hundreds of pilots; if interviewed, all would have interesting stories to tell. However, let us depart slightly from the normal view of pilots and look in some detail at just one...

Mention the word 'women' in the context of flying and many minds immediately click over to the image of 'stewardesses' - there to pander to the passenger's every whim. But this is not how it is, nor how people should think.

DAN-AIR SERVICES

One of the few women who did fly commercially before the war was Winifred Drinkwater, who obtained her pilot's licence in 1930, and was allowed to carry passengers 'for hire or reward' with the granting of her commercial licence, No.2850, in 1932. Winifred flew for John Swords Midland and Scottish Air Ferries, but she was very much an exception to the rule.

Women have been flying aircraft in large numbers since World War Two, when the Air Transport Auxiliary (ATA) was charged with *'The transport of mail, news, medical supplies, civil authorities, medical officers and ambulance work and co-operation with the Police and Fire Brigades'*. The ATA was established in 1939 under the command of Gerard d'Erlanger, who, during the late 1930s, had been a Director of British Airways. By the war's end, nearly 200 women pilots had been trained with the ATA to fly everything from light aircraft to four-engined bombers.

Even after the war, commercial flying was very much a man's world and the attitude that flying airliners was man's work' predominated. One ex-ATA pilot, with all the correct licences, skills and number of hours flown, applied to British European Airways during the early 1950s, giving just her name and initials, but without saying that she was a woman. BEA accepted her, subject to a successful medical by their doctor, and fixed up an appointment for this. When she arrived the doctor told her "...*that we are not doing stewardesses to-day!*" When she told him that she was there for a medical as a pilot, he refused and promptly sent her away. Later she got a terse letter from the airline, stating "*that it is not the policy of the company to employ women as pilots*"!

Slowly women did manage to break into commercial flying. Another ex-ATA girl, Monique Agazarian, became Managing Director and Chief Pilot of the Heathrow-based Island Air Services during the 1950s, and employed a number of women pilots, but it was still very difficult for women to break established opinions and traditions.

It was not a question of 'women's liberation', or proving that they were better than the men, although many had to do just that in order to get into the fringes of commercial aviation. It was just the desire to be granted the simple wish that if they were qualified and had the necessary experience they should be considered on equal terms when it came to filling positions on the flight deck.

For many years vehement discussions raged within the airlines on whether women were suitable for the flight deck. A popular argument adopted by many Captains within the industry was that 'women were not strong enough to handle the controls'. While in the days of the York and DC-4 this may have been partially true, as the loads on the flying controls could become very heavy at speed and

FLYING EQUALS

required a considerable effort to move them, it became a less valid viewpoint as more sophisticated aircraft fitted with power-assisted controls began to appear. Physical strength was no longer an everyday requirement, and thus the cornerstone of many objections was removed.

Captain Arthur Larkman well remembers how the first woman came to be employed by Dan-Air...

"I had completed an Instrument Rating Examiners Course at Stansted and one pilot I examined was Beryl Sanders, who was remarkably good, so I persuaded the Company to take her on as a First Officer".

There were others, as Captain Moody recalls:

"Back around 1965 (just after I got my Command) I remember flying with Jill Cazalett. There was also another lady, Claire Roberts, whose husband used to operate a DH84 Dragon on pleasure flights around the Tower at Blackpool".

In many respects Dan-Air has always been an innovative airline and in the context of employing women pilots, was one of the first in the UK so to do. Dan-Air adopted exactly the stance that was required - equality. If an applicant for a flight-deck position was qualified, medically and technically fit with the right experience then they would be considered for the post, regardless of sex. The medical given to flight deck crews is still strictly 'unisex' and makes no allowances for male or female.

With flight-bag in one hand and handbag in the other Captain Yvonne Sintes walks away from a Dan-Air Skyways 748. (Y Sintes)

Nevertheless, this was a somewhat brave step for the Company to take, for there were fears (unfounded as it turned out) about the reaction of the passengers when they heard a female voice making the *'This is your Captain speaking...'* announcements over the cabin PA!.

Dan-Air's 'First Lady Captain'

Yvonne Sintes joined Dan-Air on 1 January 1969. Yvonne's life-story (not just her career in aviation) is worthy of a book in itself and is worth looking at in some detail so as to set the scene for her time with Dan-Air.

With a life-long passion for flying, South African born Yvonne started her flying career in the early 1950s as a stewardess flying with BOAC, because of her ability to speak Spanish, on their South American routes. Working for BOAC automatically granted Yvonne membership of the Airways Flying Club, where she began to learn to fly at her own expense. She gained her Private Pilots Licence and then joined the RAF Volunteer Reserve, flying DH Chipmunks.

"Much of my instruction came from my first husband Eric Pope - he gave me an Instructors Course and I practiced on RAF aircraft until I obtained my Assistant Instructoris Rating in 1953."

DAN-AIR SERVICES

Sadly, in 1957 Eric died and, in order to bring up two young sons, Yvonne took a job as Flying Instructor at Exeter, where she helped start and run a commercial flying training school.

"Usually my career progressed with a qualified man suggesting the next stage. When I was a stewardess one of the Captains with whom I flew suggested I become an Instructor. Later someone else said they would like to see me as the first woman commercial airline pilot - and that became my ambition.

Along the way I became the first woman Air Traffic Controller with the Ministry of Aviation (there were other women controllers at non-state airfields). Of course, although the Civil Service advertised that applications would be accepted from both men and women, they never really wanted any, but as I had the right qualifications, I was accepted. I completed the training and worked at Gatwick, both on the Approach Radar and in the Tower from 1961 to the end of 1964, interspersed with a spell as Air Traffic Control Officer at Stornoway Airport in 1963.

I decided then that I desperately wanted to get back into flying, and as a number of my old pupils were now flying out of Gatwick, one invited me for a flight on a Morton Air Services Dakota. It just so happened that the Chief Training Captain was aboard that particular flight and said to my young ex-pupil that he could fly it outbound. I was then invited to sit in the left-hand seat whilst my pupil showed me something of the Dak. Much to my surprise I was asked if I wanted to fly it back! During my days with ATC I had been checked out as a Captain on the Ministry of Aviation Doves so that I could help with radar target flying at their school at Bournemouth. I had also done night flying on the Dove in my spare time at Stansted in order to obtain sufficient hours for my ATPL, so I was in practice on a similar type.

As a result I was invited to come back and help on further flights when time permitted, (I was still supernumary) doing a number of night newspaper flights until I was asked if I would like to get the Dak on my licence. I managed to get though the exam and eventually obtained a job with Mortons, who were then an associate company of British United Airlines. For the first six months I only flew freight Dakotas, but was then checked out as co-pilot on the Heron, flying passengers on their scheduled services and finally on the Dove again for executive flights.

This was during a period of industrial problems with the industry trying to get pilots to sign individual employment contracts as a result of which many of Mortons pilots joined the British Air Lines Pilots Association (BALPA). Although they wanted to support the union, none wanted the job of running the local branch, so somehow I became proposed as the Pilots Local Council (PLC) Chairman! There were numerous problems and, in the end, everyone but myself capitulated and I was left as the only one who would not sign an individual contract. So I received three months notice from Mortons to find another job! I was very concerned that my BALPA activities over the contracts - which had become well known by the airlines - would make me unemployable for a while, but I had to work somewhere.

A friend of mine, Beryl Sanders, had been flying part-time with Dan-Air and was happy there, but had to give up due to the Ministry demanding full-time pilots only. I had wandered into Dan-Air on one occasion and picked up an application form, but loyalty to Mortons forbade me completing it. But when I was given notice, I went back. I had just filled in the form when who should I bump into but the Dan-Air PLC Chairman! He asked me what I was doing, and when I told him I was looking for a job, I was whisked straight in to see Captain Bob Atkins, the Chief Pilot. Bob looked a bit shattered, but offered me a job on Daks at Bristol.

The heart-warming thing about this episode was that just after accepting the job and leaving Bob Atkins office, I met one of the ex-Morton pilots who painfully explained to me why he had to sign the personal contract. Captain Atkins must have realised what was happening and out of the corner of my eye I saw him wink at me in sympathy!

I flew the Dakota out of Bristol for six months, after which I was offered a post on the Ambassador at Newcastle, but due to the problems of seeing my children

FLYING EQUALS

Marylin Booth (right) and Deli Gray Fisk (left) on the flightdeck of a Boeing 737-200. (DAS/GMS Collection).

who were at school in the south, I was transferred to the Ambassador fleet at Gatwick. Three months later I was posted again, as a First Officer on the Comet!

The Comet IV was wonderful to fly, and I had two years on it. There was one unusual flight I remember well - it was the first time I flew the Comet with Captain Atkins. On 26 June 1970 we flew G-APDM to Akrotiri in Cyprus with an almost full passenger load. On leaving Gatwick, the undercarriage was reluctant to come up but finally obliged as we entered the 'hold' at Mayfield. On the return journey we were empty and Bob Atkins allowed me to fly it As we were approaching Athens, Air Traffic Control passed on a message from the company, telling us to divert to Ghilli. As I only knew of Gigli the opera singer, I asked for the four-letter code which gave us Izmir in Turkey. One of our other Comets had gone tech and so we transferred their full load onto ours. When it came to requesting permission to start up, we were told no, we were not allowed to start, because we had not, as yet, been given permission to land by the aviation authority! So we had to sit there quietly fuming on the tarmac for nearly another hour whilst the paperwork 'caught up' and we were officially declared 'landed' and could therefore 'depart'. We all put in 15 hours duty that day!

Then came my chance of command, flying the 748 initially out of Luton and then Gatwick. Whilst on the 748 I made a number of flights with Marilyn Booth as my First Officer - the first time on the 12th April 1975, when we flew Gatwick - Bristol - Beauvais and then back to Gatwick. Marilyn had also started off as a Stewardess, although she learned to fly and obtained her PPL in Australia. She was the first woman to be sponsored through airline training by Dan-Air. Despite any qualms that the hierarchy may have had about all-women crews, many passengers thought it was a good idea! On one occasion I took a party of ladies to Amsterdam - they had not seen my enigmatic male colleague and said to the stewardess and myself as they left the aircraft 'how nice it was to have an all woman crew'. My First Officer came up behind me in time to hear the comment and was not amused!

Of course there was prejudice shown against women pilots, but I came across very little within Dan-Air. As far as I am aware there never was an official policy about employing women pilots within the company - if they needed a pilot and a qualified woman was available, they hired her - it really was as simple as that!

In 1959 Yvonne had been the initial winner of the Brabazon of Tara Trophy for her contribution to aviation as an instructor. During 1965 she was awarded the IAOPA ATC award *'for the guiding and guarding of air safety'*, the Amelia Earhart Scholarship in 1967, the

DAN-AIR SERVICES

Sir Alan Cobham Award in 1969 and then on 20 May 1974, at the Royal Aeronautical Society in London Yvonne was presented with the Whitney Straight Award by the Princess Anne, Mrs Mark Phillips, an award that recognised the achievement and status of women in aviation. The citation read *"For exceptional courage and determination, under difficult circumstances, in the pursuit of a career with the highest qualifications in aviation"*.

Europe's first lady Captain on jets

Yvonne's promotion was certainly meteoric - much was down to her ability, but some was through being in the right place at the right time. By 1975 she had been posted again, this time as a Captain on the 1-11 fleet.

> "I moved across to the 1-11 and so became possibly the first woman in Europe - certainly the first in the United Kingdom - to command a pure-jet airliner."

After a spell of training and conversion flying, at 0914 hrs on 16 June 1975 Yvonne took the controls of G-AZED, and took off for a final line check flight with Captain Spurrell, the 1-11 Fleet Manager, for a flight from Gatwick to Heraklion, Crete, thus becoming the UKs first-ever pure jet Captain - an historic day for both Yvonne and Dan-Air, which was commented on to the passengers by Captain Spurrell.

It is crystal clear that Yvonne loves flying and she openly admits that she cannot understand why some people are scared of it, although she has always suspected that this is due to lack of information.

> "Since I started flying passengers I have always made sure they are as well informed as possible. During my time with Dan I always used to tell the Stewardesses that if they noticed someone they thought was afraid of flying to let me know. There really is nothing mysterious about flying. When we had time, often I would invite nervous passengers up to the flight-deck. After explaining to them in simple terms how it all worked, many felt much better.
>
> I could have moved onto either the Boeing 707 or the 727, but that meant I would have been away from home too much - and anyway I preferred to fly the European routes!. By now I had remarried a very simpatico Menorcean called Miguel Sintes and, due to health reasons had decided to retire to a cottage in the country in Menorca in 1980.

1-11 G-AZED, the aircraft in which Yvonne qualified as Europe's first lady jet Captain. (DAS/GMS Collection)

FLYING EQUALS

Yvonne's retirement... On the ramp at Mahon with the rest of the flight crew in front of G-BDAS after her last flight as Captain (Y Sintes)

Dan-Air's Publicity Department made much of my 'retirement', arranging plenty of press coverage for what was supposed to be my last flight - very conviently into Newcastle. But that was just for the papers. I finally retired two weeks later, after a flight I specially asked for to my beloved Menorca! On 2nd September 1980 I flew G-BDAS as Captain for the last time into Mahon and let First Officer Wilson fly it back!".

Yvonne still has many of her log-books, inummerable press cuttings and letters sent to her from happy passengers. There is one note, written hurriedly and pressed into the number one stewardess's hand after landing that always manages to raise a smile on Yvonne's face...

'Dear Madam
I may be a 'chauvinistic pig' but I take my hat off to the most skilful pilot I have ever flown with!
Yours, C. Larbie
PS - This is my first flight!'

Other females also joined Dan-Air; one such was Elizabeth Overbury (who had previously flown with Autair and Court Line).

"I am sure many people fail to understand the wonderous joy of sitting in the sharp end and seeing wonderous sights! Breaking through the overcast into brilliant sunshine with beautiful cumulus clouds all fluffy like cotton wool; a sunrise, or better still, a superb sunset with the attendent mass of colours! Seeing the Alps from 30,000 ft, Concorde climbing out through your level, or just seeing the approach lights and runway after a difficult approach...

I came into aviation through a marvellous careers teacher (this was the late 1940's and she was a very forward-thinking lady) and a determination to be a pilot. On leaving school I worked at three jobs each day Monday to Saturday and took flying lessons on Sunday. I got my Private Pilots Licence at 17 and went on from there. With the demise of Autair/Court Line in 1974, I applied to Dan-Air, but was told there were no vacancies, so took myself off to join Matabeleland Aviation in

DAN-AIR SERVICES

*Captain Elizabeth Overbury (who joined Dan-Air from Court Line in 1975) at the controls of a 748.
(DAS/GMS Collection)*

Rhodesia for a year. I was about to sign-on for another year's contract with them when Dan-Air contacted me saying that I could start on their 1-11 fleet from 4 November 1975... I was back like a shot and enjoyed the next 14 years until I retired in 1989.

With the introduction in January 1976 of the Equal Opportunities Act, which made it illegal to refuse a qualified, technically competent person employment on grounds of sex, an incident arose that was highly complimentary to Dan-Air. It then became the 'done thing' for airlines to proudly display that they had female pilots. The problem was that the national carrier had no female flight deck members! To rectify this, they quickly made 'an offer they could not refuse' to a number of Dan-Air female pilots!

Yvonne was soon joined by others, for by now the mould of 'male only' flight-deck crews had been well and truly broken. Delphine (known to all as 'Deli') Gray Fisk from Skyways, Val Stanley, Shereen McVicar... all these and a number of other ladies found their places on the flight-decks with Dan-Air.

So that concludes our look at some of Dan-Air's pilots; indeed, 'Flying Equals' all! Perhaps we should let Yvonne sum up her flying career on behalf of all Dan-Air's pilots - both male and female. This can be best done by a small paragraph taken from a letter from Yvonne to the author...

"I am truly grateful to Dan-Air for taking me on as a pilot, enabling me to earn a living doing the job which I loved. They completed a dream"

DAN-AIR SERVICES

CHAPTER EIGHT
New aircraft, new routes.

1978 saw the airline celebrating 25 years of operations - and they did it in fine style, carrying over four million passengers! It was also the year in which Dan-Air was second only to British Airways in terms of numbers of passengers carried by a British airline. The figures published in *CAA Annual Statistics* (CAP 406) make interesting reading, especially the section detailing costs per seat kilometre. British Airways and British Caledonian (with a mixture of routes) turned in a figure of 2.7p/seat-km, whilst Dan-Air with a mix of scheduled and charter routes returned 1.24p/seat-km. Lowest of all (because of the high load factors generated by its all-charter operaton) was Britannia Airways, with 0.95p/seat-km. Nevertheless, Dan-Air's figure was impressive.

All Dan-Air's equipment was organised on 'cost-centre' lines according to aircraft type. Each fleet was financially separate and accountable to Head Office both operationally and technically. Thus, whilst all selling was centralised, maintenance of aircraft by Dan-Air Engineering was chargable to the appropriate fleet.

There were also tentative thoughts about expanding into larger aircraft types, though anything above 200-seat capacity was considered too large. Frank Horridge (then Managing Director) told ***Flight*** that the airline had looked at the Boeing 757 among other possibilities for long-term IT business but that the cargo space would be wasted in the IT market. As Martin O'Regan, the Financial Director, said at the time...

> "It is one thing having an aircraft that can offer low seat-mile costs. but unless the aircraft is filled, then its potential connot be realised. We have no plans to operate wide-bodied aircraft, but when the market was right it would be logical to add some to our range".

Edition One of Dan-Air's In Flight' magazine dated January 1978.

The UK-generated IT business formed a little under half of Dan-Air's business, while a further 20% came from overseas. The scheduled service and oil-related charters accounted for a further 20%, the remainder coming from miscellaneous activities, including the engineering division.

The scheduled services had some useful routes and was expanding at a rate that could be carefully monitored. However, the then current round of route applications were regarded as 'nonsense' by Mr Newman, the airline's Chairman. *"Even if they were awarded, the countries concerned would not grant traffic rights"* so Dan-Air objected in writing and withdrew from the proceedings. Mr Newman also thought that European de-regulation was not the answer to air-transport problems. *"Our authorities should benefit from the experience of the USA, where* profits have taken a downturn. In a business as sophisticated as air transport, some regulation is necessary in the interest of the consumer and the airline".

The Silver Jubilee year saw a number of interesting flights taking place. One of the Company's 1-11s was chartered to carry the Chinese Vice-President, Mr Wang Chen, and his delegation during a visit to the United Kingdom, and the year also saw the last trans-Atlantic flight, the Company preferring instead to concentrate on developing the European market.

In preparation for the IT affinity flights to Hong Kong from Gatwick, permission for which had been granted in 1974, a number of Dan-Air cabin staff were introduced in January 1978 to the mysteries of Far East cuisine by Far East Travel Centre, to whom the flights were chartered. The service was to commence on 14 February and passengers on these flights would be served both English and Chinese cuisine, so the cabin staff needed to know all about the food and how to handle chopsticks!

However, 'the writing was on the wall' for the airline's long-haul operation; two 707s were sold off at a profit and a third returned to its owner. A further 707 was retired at the end of the 1978 summer season and the final machine was leased to another carrier.

The final year of the decade saw a general up-turn in business, with turnover and profitability rising for Dan-Air, even if the number of passengers and distance travelled had declined. This was caused by the termination of the long-haul side of the airline's activities, which brought about a more concentrated sphere of

Eva Lam of the Far East Travel Centre (in gown and mortarboard) tries to explain Chinese and Cantonese food to (L to R) Sue Evans, Lynne Boreham, Velma Sharma and Sue Hale. Dan-Air was starting to operate a number of Far East charters for FETC! (DAS/GMS Collection).

NEW AIRCRAFT, NEW ROUTES

operations, and in order to become even more efficient, a new type of aircraft was needed.

As early as 1972 Dan-Air had begun to search for a replacement for its fuel-thirsty Comet fleet. A report appeared in the 6 April edition of *Flight* stating that they were considering ordering '*up to six Boeing 737-200 series aircraft in a 140-seat configuration, which would be used primarily for IT services to the Mediterranian area but that final details have not been decided upon*'. After studying performance figures, eventually the Boeing 727 was decided upon as the logical and worthy successor to the Comet, the first three being -100 series ex-Japan Airlines.

The Boeing 727 design originated as a paper study in February 1956, at a time when it appeared that the British De Havilland Comet and the French Sud-Aviation Caravelle were going to dominate the American domestic market. Boeing Aircraft Company of America decided to go for a genuine short field performance aircraft in order for it to fit into the many smaller airfields, such as New York's La Guardia and Germany's Berlin Tempelhof, into which no other jet of this size could operate.

Over 150 design studies were made before the final configuration was fixed on 18 September 1959. From there the programme moved swiftly forward, the prototype taking to the air on 9 February 1963. In order to achieve the required short-field take-off performance, complex high-lift leading and trailing edge slots and flaps were installed on the wing.

Thoughts were also given to achieving the maximum turn-around speeds with the minimum amount of ground-handling equipment. An AirResearch auxiliary power unit to provide air conditioning, ground power and compressed air for engine starts was fitted, along with an integral aft stairway and a forward cabin door, fitted with optional air-stairs for ease of passenger boarding. Boeing were soon able to offer several major variants, including a major re-design - the series 200 - with a 20 foot longer fuselage to enable up to 163 passengers to be carried, increased to 173 passengers at a reduced seat pitch.

G-BAJW retracts its undercarriage as it climbs away from Manchester. (GMS/DAS Collection)

DAN-AIR SERVICES

Before the machines obtained by Dan-Air could enter service, much administration and engineering work had to be done, including certification for British operation, work that was undertaken at the manufacturer's plant. It may seem surprising considering the world-wide popularity of Boeing's tri-jet that the initial three examples obtained by the Company were the first of the type to be granted a British Certificate of Airworthiness.

Each of the three ex-Japan Airlines -100 series machines was ferried by Japanese crews across the Pacific to Boeing's plant. Unfortunately these aircraft were not equipped with high frequency (HF) radio, so each machine made an epic trans-Pacific 'buddy flight' alongside a scheduled 747 flight from Tokyo to Los Angeles, the 727 landing at Wake Island, refuelling and waiting 24 hours until the next scheduled 747 overflew Wake. It then stayed in contact with this 747 until reaching Honolulu where another 24 hour stop was made. The journey progressed in this manner to landfall at Los Angeles from where the crew could head north for Seattle. The first machine arrived at Boeing's factory during October 1972, other machines following at monthly intervals.

The major task to certificate the aircraft for British use was the design and installation of a full stall protection system, consisting of a stick pusher, a stick nudger and an independent stick shaker for each pilot. Two machines were fitted with extra fuselage-mounted fuel tankage in order to allow them to fly from Berlin to Tenerife non-stop with a full passenger load.

Dan-Air also required the passenger-carrying capability to be increased from 131 to 153 seats, something that had never been done before. Boeing installed additional emergency exits in the rear fuselage and a whole series of trials was conducted to ensure quick and easy evacuation of the aircraft with a full passenger load in the event of a problem.

With these aircraft entering service under contract for Lunn-Poly, Global, Clarksons and Sun-Med, the Company could now offer the tour companies pure-jet aircraft with a wide range of seat options from the smallest, the BAC 1-11 with 89 seats, through the Comet with 119, the Boeing 727 with 153 to the Boeing 707 with 189 passenger seats to suit any tour company's specific requirements.

The first machine to arrive at Gatwick registered to Dan-Air was G-BAEF, still in full Japanese Air Lines colours, but wearing its British registration. However, the machine was soon flown to Lasham for re-painting in Dan-Air colours.

Hand-in-hand with the acquisition of the 727-100 aircraft, a 727-100 series simulator was obtained for flight deck crew training. This was an analogue computer-controlled device with only a limited degree of movement and no display in the flight deck

Edition Two of Dan-Air's 'In Flight' magazine for the summer of 1978.

NEW AIRCRAFT, NEW ROUTES

With ground power and air-start carts plugged in 727 G-BAJW is seen at Gatwick preparing to depart on another trip. (DAS/GMS Collection)

windows, so its use was limited to night and instrument flying. Nevertheless, it was a vitally important instrument for crew training.

Later, a 727 series 200 six-axis simulator was installed at Gatwick, with full movement and electronically generated images on the front flight deck windows, allowing a much greater degree of realism to be simulated. Both devices were used not only by Dan-Air - any excess capacity was 'sold' to other airlines.

One of the longer-range Boeing 727s was permanently based in Berlin, for in early 1972 Dan-Air took over the Orientair contracts. This Inclusive Tour company had been formed by Captain Lockwood (formerly of Channel Airways) who, following that company's demise, bought a BAC 1-11 and obtained a number of contracts from West German tour operators. The plans ran into financial difficulties and Dan-Air took over the programme.

The airline worked the 727 hard, as a typical month in the flying life of Captain Robert Burgess, a member of the 727 fleet, shows:

Edition Three of Dan-Air's In Flight magazine for the winter of 1978/9.

"The first day of October 1988 found myself and my crew taking a leisurely breakfast at the Holiday Inn in Bristol. 1988 had been an exceptionally busy year for us all, with the media giving the industry a tough time about over-crowded airports and assorted Air Traffic Control delays. For our part we had all flown a busy summer programme to all corners of Europe and were hoping that October would be just a little bit quieter.

From Bristol we went to Palma with 202 passengers (Pax) and crew aboard G-BHVT, continuing on back to our base at Gatwick with 196 on board. The next two days saw trips from Gatwick to Munich on G-BHNE and Menorca on G-BCDA. The 6th was Gatwick to Madrid and back using G-BAJW, followed by a taxi ride to Great Dunmow in Essex to cover a series of flights from Stansted. As a crew we all went out to a local Indian restaurant and, upon returning to the Hotel, a message awaited me. The weather forecast showed a severe depression tracking over Northern Ireland and Central Scotland - glad we are flying south I thought.

My telephone call to Dan-Air Operations to pick up the message produced the request *'We want you to position across to Belfast in the morning and take 130 V.I.Ps to Nice and then position empty to Stansted'*. So much for flying south away from the bad weather! Next morning we parked G-BAEF on the rain-swept, windswept apron at Belfast International Airport. The weather was so bad we decided to load our guests via the ventral stairway at the rear - we dared not open the front, main passenger door! The flight to Nice took just 2 hours 9 minutes - this is always a memorable flight, by virtue of the descent on the ILS from St Tropez, past Cannes and Frejus then circling out over the bay to land on the south-west runway.

DAN-AIR SERVICES

Our agent at Nice as usual was pleased to see us, hurrying aboard with a Telex - new orders. Dan-Air had booked a sub-charter from Air Europe to take 140 Pax from Ibiza to Gatwick. So, off to Ibiza in 1 hour 18 and then on to Gatwick.

Stansted - Palma - Stansted using G-BAEF again, then after days off, a trip to the island of Crete with G-BAJW. This island is so mountainous it seems impossible that there is room for an airport - nevertheless, two commercial airports thrive there; Heraklion, just to the east of the capital and Khania, far to the west near Suda Bay, the site of the savage battle for the islands in 1941.

12th October and we're off from Gatwick to Isreal aboard G-NROA, a flight that shows the 727 at its best. Always a maximum take-off from Gatwick - an all-up weight of 91.5 tons with usually 28.5 tons of fuel aboard. Tel Aviv takes about 4.5 hours and it is always fascinating to watch the island of Cyprus on one side of the radar screen and the Egyptian coast on the other give way to the coast of Israel with, just a short distance beyond, the radar picture of the Dead Sea valley. Such a tiny country. The return flight to Gatwick again requires maximum take-off weight and a stepped climb. Our flight time home was 4 hours 54 minutes, a round trip total of 9 hours 26 minutes.

I took a week's leave in a cottage on the Isle of Wight - pilots have holidays too! - before I was off to Berlin (as a passenger this time) to operate as crew on our German charter programme.

October 24th saw me on a Berlin to Las Palmas in the Canary Isles using G-BHNE. This flight represented the ultimate range of our 727 with a full load of around 180 pax. At nearly 2,200 miles it is further than Shannon in Eire to Gander in eastern Canada, the shortest distance across the Atlantic! Our flight down to the Canaries takes 5 hours 6 minutes, routing down the centre corridor from Berlin, across the border at Wolfsburg, south over Frankfurt and Zurich. South past Lake Geneva and Mont Blanc, across southern France to Toulouse, across the Pyrenees, over Madrid, then Seville and a magnificent view of the straights of Gibraltar before routing direct to the Canaries.

The homeward route to Berlin was via northern Portugal and the Bay of Biscay to pick up tail-winds, then over the top of Paris towards Cologne and Hanover before descending to low-level below 10,000 feet for the flight along the centre corridor and into Berlin.

On the 25th it was Berlin to Montasir in Tunisia and return, repeated again on the 26th. The 27th was a day off in Berlin before travelling home that evening, so after breakfast I went on the S bahn to Wannsee to take the steamer up the Havel to Alt Tegel for a late lunch.

The next day was Gatwick to Menorca again with G-BCDA and on the last day of the month it was Gatwick to Faro in Portugal aboard G-BHNF. This is another airport that has expanded rapidly in recent years to accommodate the massive increase in people travelling to the Algarve.

All in all, a busy but typical month on the 727 fleet. 74 hours 43 minutes flown, to 16 different airports in 10 countries and 3 continents. We made 28 landings and take-offs and carried 4,500 passengers - oh, and I had a week off as well!'

'In Flight' magazine for the summer of 1979

Showing its sleek lines, here G-BAFZ waits on the ramp at Gatwick. Note that the pristine condition is spoilt by the paint that has peeled off the nose area!
(DAS/GMS Collection)

124

NEW AIRCRAFT, NEW ROUTES

Potential delays and the VIP treatment...

Delays (for whatever reason) can and do occur on any airline, let alone one of the size of Dan-Air. Many of the airline's charterers keep detailed records to see how the listed departure times, prepared many months in advance, compare with the actual. These figures, when viewed over a period of a season, reveal some startling information. The following is from data provided by the charterers Owners Abroad to Dan-Air, and related to all flights featured for the 1983 season of the OSL and Wings brochures:

	All Charters	Dan-Air
Total Flights	11424	1078
Flights departing within 30 minutes	84.5%	87.8%
Flights delayed 30-60 minutes	8.3%	7.0%
Flights delayed 1-2 hours	4.4%	2.8%
Flights delayed 2-3 hours	1.6%	0.9%
Flights delayed over 4 hours	0.8%	0.7%

In 'Flight' magazine for the winter of 1979/80.

As can be seen, most flights got away within half an hour of their scheduled time. Dan-Air out-performed the average across the board, in some cases by a handsome margin!

Passengers also have some interesting tales to tell about potential delays, flying in the 727, and how the airline attempts to minimise any inconvenience, even if normal procedures do tend to go out of the window. Simon and Abbi Peters recall a flight to one of the lesser Greek Islands...

A famous and quite beautiful photograph that captures G-BHNF as it taxies out for another flight from Gatwick. (DAS/GMS Collection)

"We had booked a holiday for June 1981 with Sun Med Holidays to the Greek island of Santorini. At that time there were no flights direct from the UK so we had

125

DAN-AIR SERVICES

an evening Dan-Air 727 jet from Gatwick into Athens and then a domestic flight by Olympic to the island. On coming back, we got to Athens Airport well in advance, but were informed by the Sun Med tour representative that although we were supposed to leave around midnight, there was at least an eight hour delay in getting back to Gatwick due to Air Traffic Control problems.

There were hundreds of people milling around the terminal, for it was not only our flight that was delayed, so we found a space on the floor to sit with our cases and wait. After a couple of hours the PA system called our flight number, announcing that all passengers were to go to a particular check-in desk immediately. It seemed that somehow an earlier take-off had been arranged and now the flight was leaving in about 20 minutes! Our cases were taken in and in no time at all it seemed we were rushed through the departure lounge and onto the aircraft.

At the top of the rear stairs on the aircraft we were met by a stewardess who was telling everyone to sit in the first seat that they came to - we would be able to sort everything out when we were airborne. If there was any delay now, we would have to wait for eight hours!

There were other girls at intervals down the cabin making sure that we did what they wished. It was the quickest we have ever seen an aircraft loaded; there was none of the usual messing around by the passengers, the cabin staff were helpful and polite, but very firm - just put your hand luggage in the nearest overhead bin, sit down and strap in! In no time at all (it seemed) the doors were shut and we were taxying towards the runway. Safety announcements were made and we were off!. We both got the impression that the crew did not want to wait around any more than we did and they did all they could to make sure we hit that take-off time!"

In 'Flight' magazine for the summer of 1980

But it is not always stories of delays and frustration - sometimes the passengers can find themselves inadvertently getting the full V.I.P treatment! At the start and end of the tourist season, aircraft often flew one way empty, for there were no passengers to carry on the other leg of the flight...

Micheal and Vera Garratt were set to return from Rhodes in the late afternoon of 11 April 1984 after celebrating their 25th wedding anniversary on the island. On arrival at the airport they discovered that they were the only passengers there!

"At first we thought the airport was on strike, but as strange as it may seem, Manchester United were playing Juventus of Turin in the first leg of the semi-final of the European Cup Winners Cup at Old Trafford. The match was being televised 'live' and most of Rhodes were United supporters!

We waited at the check-in desk and eventually one of Olympic Airways staff came over and told us that our Dan-Air flight to Manchester was on schedule and that we were the only people on it. After coffee in the empty cafeteria, we were called to Customs where we were ushered through into the empty departure lounge very quickly. The shops there were closed, but the Manager came over and congratulated us on being the only people catching the flight, then opened the shops especially for us, the staff appearing from nowhere. However, the bar was not open, so I had to hop back through Customs to obtain a couple of cans of beer and a Coke, the Customs Officers turning their heads the other way as I ran back. Within minutes the loudspeakers announced the early departure of our flight due to lack of customers and that '...all passengers are instructed to go to Gate 9 where transport is waiting to take you to the aircraft'. We went down the ramp to the bus where a full contingent of airport staff was waiting to help us with our luggage.

On boarding the aircraft (a Boeing 727-100) we were met by a full set of cabin staff and my wife and I were given seats in the front of the cabin with plenty of leg-room and comfort. We could not really believe what was happening - we were going to fly back to the UK in an aircraft completely to ourselves, being looked after by a full crew! On take-off, the cabin staff ran through the full flight safety instructions as if the aircraft was fully loaded and, we were informed, because the circumstances of the flight had been changed, we would be flying at 40,000ft. I was

In 'Flight' magazine for the winter of 1980/1

126

NEW AIRCRAFT, NEW ROUTES

Micheal and Vera Garratt, the only passengers aboard this 727 heading back from Rhodes
(M & V Garrett)

allowed to go into the flight deck after take-off and for over an hour watched the sun setting on the horizon - it appeared stationary for some time.

In-flight service could only be described as 'superb'; the drinks came faster than normal and we were told that as they had '57 cottage pie meals on board, if we wanted more for the meal, we only had to ask! The cabin staff were great fun and very friendly - they went out of their way to make our journey a pleasure. At Manchester we were given the VIP treatment by being invited to use the Dan-Air crew-bus from the aircraft to Customs and one of the stewardesses offered to take us back to Bury on her way home to Burnley!

Working away from the UK...

Of all the many airports that Dan-Air operated into, a number had small teams of employees permanently based to provide a link between on one hand, the visiting aircraft, its crew and passengers and, on the other the operators and contractors of the airport. The aircraft itself may have been 'turned around' - unloaded, refuelled, cleaned, re-catered and passengers' baggage put aboard - by handling agents, but this was overseen and co-ordinated by Dan-Air's Airport or Station Manager and a team of Traffic Officers.

A good representative example of this kind of outstation operation could be found at the Aeropuerto De Menorca. Here

Passengers de-plane from 727Alpha November into the sunshine of Malta.
(DAS/GMS Collections)

DAN-AIR SERVICES

Margaret Gray was the Airport Manager (a title that could be confusing at times - she was often asked whether she ran the Airport itself) with a small team of Traffic Officers - sisters Kathy and Diana Weller, Andrea Collier and Frank Ellis. During 1992 the team, working during the peak summer months on early and late shifts, looked after nearly 40 flights per week. Margaret explains just what went into the position:

"We existed to provide a link between the crews and the Airport and between the tour operators and passengers and the Airport. I attended Airport Committee meetings, laised with Aviaco, our handling agents, worked closely with the tour operator charterers to ensure that all were informed as much as possible. Not only did we have our own flights from the UK to consider, but also those of Monarch, Airtours, InterEuropean, Braathens Safe and Scanair. We had to negotiate return slots with Palma Control, assist the crews with flight planning their return route and sort out any problems."

In 'Flight' magazine for the summer of 1981

As can be seen, it was not a just a question of 'catch 'em, turn 'em round and throw em' back!' Particularly at the beginning and end of the summer season the tour operators consolidate, or amalgamate a number of their charter flights into one in order to make the most efficient use of aircraft. If this was not handled correctly, it was all too easy for the passengers to perceive this re-arrangement of their travel plans as being caused by the airline, not the tour operator.

'I tried to work closely with the tour operators both on the island and back at their headquarters to ensure that the travelling public were as well informed as possible and that they got a consistent story. We found that people were far more happy and prepared to accept the truth sooner than excuses.'

Nevertheless, there were times when the public could make our jobs difficult. Nowadays, people are much more used to travelling by air and many never give it a second thought. Unfortunately, a number have become so blasé that they think they can use an airline just like they would a bus service, but it's much more complex than that. Normally we never had this problem with tour operators who arranged ground transport to the airport from their holiday accomodation with plenty of time to spare. But we got a high percentage of seat-only flights into Menorca - villa, apartment ,yacht owners and the like - and some turned up to catch a flight only minutes before it was due to depart. We tried our best to get them through the formalities and aboard their aircraft, but we could only do so much!

Nevertheless, it was an absolutely facinating job with never a dull moment - I would not have swapped it for the world!'

Passengers are escorted out to and shown aboard 1-11 'CK by Dan-Air's outstation staff. (DAS/GMS Collection)

128

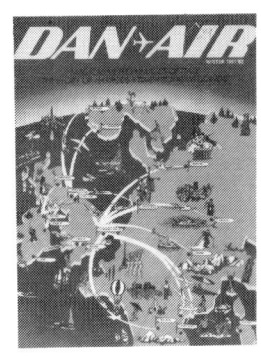

NEW AIRCRAFT, NEW ROUTES

It was not only the ground staff that found themselves living (and working) away from the UK for extended periods of time. For at least nine years Dan-Air operated 'The Royal Flight' on behalf of the Sheikh of Qatar who previously had a Boeing 707, but replaced it with a brand-new 727, equipped very luxuriously to his own requirements. Two full crews were involved with this long-term detatchment and the aircraft was also maintained by Dan-Air Engineering.

Tragically, one example, G-BDAN, a series 100 aircraft and the fleet flagship, crashed into a mountainside on Tenerife and was destroyed on 25 April 1980 with the loss of all aboard.

Despite some of the set-backs, the 727 proved to be useful in service. Although the type was far more fuel-efficient than the Comet which it replaced, the ever-rising cost of fuel was becoming more noticable on the Company's balance sheet. For instance, in 1980 Dan-Air spent approximately £44 million on this item alone, an increase of 42% on the previous year!

In 'Flight' magazine for the winter of 1981/2.

Hard work and lighter moments...

In the spring of 1985 rumours abounded that there was to be yet another season of delays and strikes. Ground Traffic services started to prepare for a possible strike at Faro in Portugal and called to volunteers to come forward to fly out to handle all aspects of ground handling. The response was overwhelming and from these a team of 20 was selected. Tim Bell and Rupert Batstone formed the advance party and were despatched on 22 May. That evening, thirteen others flew to Birmingham to catch the first flight to Faro at 0445hrs the next morning.

Of the 53 planned aircraft movements at Faro, Dan-Air was to operate 15. Upon arrival from Birmingham it was obvious that the strike was indeed on - nothing moved. Loaders stood around and the steps remained untouched; the departure lounge was full to

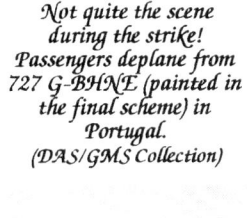

Not quite the scene during the strike! Passengers deplane from 727 G-BHNE (painted in the final scheme) in Portugal. (DAS/GMS Collection)

DAN-AIR SERVICES

bursting with hundreds outside unable to get in. Only two check-in staff were working and there were no baggage handlers!

The only practical solution was gate check-ins, which were finally negotiated with the security officers, but inevitably there was some harassment from union officials, and others who were confiscating boarding cards from some passengers! Brian Hare donned his 'negotiating hat' and soon had a workable solution which, although not totally practicable, got around the problem with some clever footwork by all concerned - a 'cat and mouse' game that was helped by the fact that the ground crews' uniforms blended in with those of the aircraft crews!

The ramp at Faro soon filled up with arriving aircraft. Passengers had to carry their own baggage both ways, but many passengers 'mucked in' with Dan-Air staff and operating crews to onload and offload baggage. An Ops Control was set up in the terminal to keep communications open with Gatwick and overall the team worked for 27 hours solid, loading, unloading, checking in, load control, start-up...

During that time all 15 aircraft were turned round; at the peak six aircraft were being worked on simultaneously. All passengers were loaded onto the correct flights, nobody (and no baggage) was left over at the end of the day and all this was achieved without the help of a terminal public address system. Ground Services 'A' team won through again!

'In Flight' magazine for the summer of 1982.

Taken during a training detail, this photograph shows the cockpit of 727-100 G-BFGN. The Flight Engineer sits sideways facing his panel. (DAS/GMS Collection).

NEW AIRCRAFT, NEW ROUTES

727 G-BAFZ in the company of 1-11 G-BCWA on the ramp at Alghero. (DAS/GMS Collection)

'In Flight' magazine for the winter of 1982/3.

That may have been sheer hard work for an extended period, with the only humour coming from a sense of adversity, but there were times when the fun came much more easily...

On 19 November 1988 a group of passengers boarded a 727 at Gatwick, bound for Alicante. Little did they realise that this was not going to be a normal flight, and the returning passengers would also be in for something of a surprise! The aircraft's crew - Captain Mike Marson, First Officer Terry McHattie, Engineering Officer Brad Bradley and cabin crew Barbra Friedrichs, Karen Muggeridge, Maxine Mitchell, Brad Mullins, Abbey Smith and Sue Wykes, were set to make sure that this would be a flight that both groups remembered, for they were intending to raise money for the BBC's *'Children in Need'* appeal.

After take-off Mike Marson masterminded a grand auction over the cabin PA of just about everything that could be moved! His own hat went for £75, stewards' tie pins were sold for a fiver each, cabin staff 'wings' were sold for £5 each. A seat on the flight deck to watch the landing reached £50, while just a look on the flight deck raised £5. Then Dan-Air models were sold. There was even an offer to buy the Captain himself, which was declined!

By the time the 727 returned to Gatwick, *'Children in Need'* were better off by the tune of £1,000, which included the entire crew's flight-pay and a final topping up to reach the magical figure! Other sections of the Company also did their bit: for example, the Inverness base 'relieved' their passengers of £571.65!

This type of fund raising was not confined to *'Children in Need'*. In March 1988 a 727 under the command of Captain Tony Anderson linked up with BBC Radio One's Simon Bates to claim a world altitude record on behalf of their passengers for wearing *'Comic Relief'* red noses! Cabin staff Sue Adams and Yasmin Sells gave up their day off to dress in clown's costumes and sold certificates during the flight to commemorate the record! Meanwhile, in Aberdeen, Captain Bryn Wayt and his crew sold a

DAN-AIR SERVICES

haggis at 33,000 feet over Carlisle and a group of Satoil employees bought the radio call-sign for the flight! Like Captain Anderson, Capt Wayt claimed a world record for his haggis auction!

With the gradual phasing out of the 727 fleet in the early 1990s the -100 series simulator housed at Horsham was also redundant. It was not scrapped however, but was donated by the Company to the Bristol University Department of Aerodynamics, where it was converted to electronic control and integrated into degree training for people coming into aviation, thus providing a continued benefit to British aviation.

Boeing's 'baby' jet

On the same day that Gatwick Airport was witnessing an emotional farewell to the world's first commercial jet airliner - the fabulous De Havilland Comet - a small, stubby twin jet quietly landed almost unnoticed. The arrival of the first Boeing 737 to appear in Dan-Air colours (a 269 series machine obtained on lease from Maersk Air) may have been a low-key affair, but the date, 9 November 1980, was to be a vitally important one for the Company's future.

The 737 had became increasingly popular with many charter airlines within the United Kingdom from as early as 1968, when Dan-Air's competitor for much of the holiday traffic, Britannia Airways, obtained the first of many of the type.

Boeing's superb 'baby' twin-jet - the 737 series - was the smallest of the 'family' of jet aircraft that started in the mid-1950s with the 707 and of which the largest is the 747. The 737 had been designed in the 1960s as a short-haul airliner to compete on broad terms with the BAC 1-11 and the Douglas DC-9. However, unlike its rivals, Boeing decided to keep the six-abreast seating already established with the 707 and 727 and also to retain the underwing engine layout of its 'big brother' the 707, with which it shared a good deal of

The special edition of 'In Flight' magazine for the summer of 1983 that celebrated 30 years of Dan-Air.

G-BICV, Dan-Air's first Boeing 737 after its repainting in Company colours. (DAS/GMS Collection)

NEW AIRCRAFT, NEW ROUTES

DAN-AIR
SYSTEM TIMETABLE

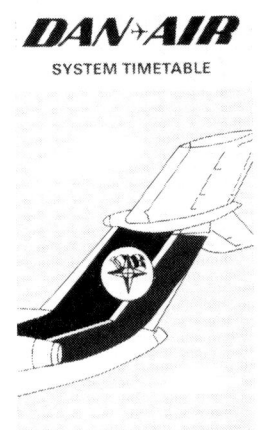

WINTER 1982/3
Valid
24 OCTOBER-26 MARCH

*The winter 1982/3 timetable.
(24 Oct - 26 March)*

commonality. A pair of Pratt and Whitney JT-8 jet engines of assorted marks were thus slung under the wings.

The launch order for the 737-100 series came from the German airline Lufthansa, who specified seating for 100 passengers. Boeing soon discovered that there existed a greater market for a slightly larger aircraft that broadly retained the original 737 layout. The 100 series first flew during May 1967, followed by the first of the 200 series machines on 21 December of the same year. Later, advanced models of the 200 series appeared, powered by turbo-fan versions of the JT-8, installed in 'Quiet Nacelles' in order to meet the USA Federal Aviation Authority's increasingly tight controls on noise regulations. Dan-Air may have been one of the later customers for the 'baby Boeing', but the type soon found success on many of its routes.

Inverness to London Heathrow

Nowhere on Dan-Air's network can offer such a challenging environment as the Highlands of Scotland and nowhere on the airline's services is the senior stewardess so well-known to the regular passengers - nor her services so much appreciated - as on the Company's Inverness - Heathrow route. Such was the success of this service that, although initially operated with 1-11s, the passenger levels soon rose to such a peak that the 737 was introduced.

This service came about when, on 19 January 1983 Dan-Air received the new route licence - the first time the airline had flown from this new Scottish destination and the first time into London's Heathrow Airport!

Maurene Perera was Dan-Air's base stewardess at Inverness Airport, responsible for the selection, training and organisation of

*1-11 G-BJMV re-fuels at Inverness whilst its sister G-ATPK taxies out for another service to London Heathrow.
(DAS/GMS Collection)*

DAN-AIR SERVICES

the stewards and stewardesses who look after the travellers on Dan-Air's three times daily Boeing 737 flights to London's Heathrow Airport, and return.

True, the overall Dan-Air standards on these services were prescribed from the airline's headquarters at Horley and were intended to be the same as those adopted all over the domestic UK routes, but, well, the Highlands of Scotland were a law unto themselves, and 'Mo', as she was affectionately known by all, applied just a few extra touches that the Highlanders (and many others) appreciated.

For instance, Mo, who was appointed the base stewardess when Dan-Air started the route, always tried to include local Scottish fare in the make-up of the meals - marmalades and jams from Moray, smoked and fresh salmon from Inverness, seafood from the West Highlands, venison from Ross-shire and products from the local creameries - not just best quality milk and fresh cream, but special Scottish cheeses as well.

To say that the regular passengers appreciated the on-board service more on these flights is an understatement, for the first departure south from Inverness in the morning was at 06.50, and some of the passengers drove over 100 miles to reach the airport in time, through rain, ice, snow or fog across the Highlands or along the Great Glen. They would probably not have had any breakfast before leaving home, and to be able to unwind in warmth and comfort on the 80 minute flight to London, and be well-fed and attentively looked after created a great affection for 'their wee airline', as they liked to think of Dan-Air. Many passengers

The summer 1983 timetable. (27 March - 22 Oct)

The place is Inverness Airport, the date December 1983 - the occasion is the naming ceremony of 1-11 G-BDAE 'The Highlander'. Mo Perera is second from the bottom of the stairs. (DAS/GMS Collection)

NEW AIRCRAFT, NEW ROUTES

returned the same night, and faced a long drive home in the dark, so equally appreciated a good dinner on the flight north.

Mo should know all about these things, for she is Scottish herself, being born not far from Glasgow, where she first started flying for Dan-Air in the 1970s. In those days she was recruited as one of the four-strong team of stewardesses flying holiday charters from Glasgow Airport to the sun-spots of Europe.

'In Flight' magazine for the winter of 1983/4.

"Business was strong enough to keep our charters operating throughout the winter as well, so I was in permanent employment. After a few years of flying the Comet out of Abbotsinch, the base was closed down for the winter, so I had to commute daily by car all the way to Edinburgh to operate the winter charters there.

After a year they re-opened the Glasgow base, and besides Boeing 727 charters to the sun, we now had a new programme - the HS 748 flights to Sumburgh. I alternated between the types of flight. Eventually, the Sullom Voe terminal was completed and the Dan-Air activity at Glasgow once again declined. I was the junior of the two stewardesses left there and just when it looked as if I would have to leave or move south, the Inverness - London service was taken over by Dan-Air from the national carrier and I was offered and took the job of base stewardess at Inverness' Dalcross Airport.

This meant moving my family up to Nairnshire, but it is something I have never regretted - the quality of life in the Highlands is unsurpassable"

The sudden take-over of the route from the national carrier meant that Mo had to recruit another 13 stewardesses in addition to herself, and so to start the Highlands service, six of the previous summer's joiners at Gatwick volunteered to go north to Inverness to become 'No.1 stewardesses' - that is, in charge of each flight's cabin crew - and another seven juniors were recruited locally. Thus Mo started the twice daily service with a team of 14, in which she had to play a full part herself as a No. 1 in addition to running the administrative side of the base.

The route had previously been a loss-maker, but a sales representative was appointed in Inverness and a major sales campaign launched detailing the changes Dan-Air were to bring to the route. By basing an aircraft in Inverness, both northern and southern departing passengers would benefit from realistic departure times. To fly there and back in a day with sufficient working time in between and receive a good cabin service would show that Dan-Air could make the route a success.

'In Flight' magazine for the summer of 1984.

The airline worked swiftly, so that within two months of being granted the licence, the first scheduled service was flown on Sunday 27 March. Although only intended as an aircraft positioning flight from Gatwick, due to the tremendous interest shown, it was changed to become revenue earning. Never one to miss out on the Scottish connection, the airline took part in the 50th anniversary of flying from Inverness, organised pleasure flights for school-children over Loch Ness and even served a breakfast of grouse on the 'Glorious Twelfth'! Bill Fawcus, who made regular use of the 0650hr INV-LHR service, remembers the first 'grouse breakfast' very well.

135

DAN-AIR SERVICES

*Grouse aboard a company 1-11!
Left to right: Mo Perera with Holly Mackintosh, Debbie Ferguson, Sue Nelson and Willie Macleod and Kevin Reid (a chef and Assistant Manager from the Golf View Hotel)
(DAS/GMS Collection)*

"...I was surprised when I arrived at Dalcross on the morning in question to be met by the skirl of the pipes and a barrage of cameras. It turned out that this was not because they had heard I was travelling, but because we were going to be treated to the first grouse of the season for breakfast!

Evidently the birds had been bagged on the Earl of Cawdor's estate by the Earl himself at first light (about 4.0am) and rushed back amidst a trail of feathers as they were plucked on the way to be prepared by the chef of the Golf View Hotel at Nairn who used pre-heated ovens for a quick roast before the birds were taken on to Inverness. The flight left on time and I was treated to a very tasty morsel. This was marred only by the fact that I was sitting next to a LBC Radio reporter who insisted on thrusting a microphone into my face just as I was enjoying my first mouthful to ask my impressions!

This man had been up on the moors to witness the whole event and had the tables turned on him when we landed at Heathrow. He had to give a live interview responding to questions from the studio and was required to do some quick guessing when asked why the 12 of August!"

The team at Inverness grew as both the frequency and size of aircraft increased, to an average of 25 (latterly including Mo and a few male stewards, most of whom had aspirations of eventually gaining their wings and occupying the flight-deck seats!). All the cabin crew were recruited locally and their soft Scottish accents went down well with the passengers, not to mention their untiring efforts and attentiveness!

Flying regularly Mo has, over the years of the route's existence, came to know thousands of the regular travellers who, on this lifeline from the top of Scotland, included a handful of M.P.s, Scottish Office officials, aristocrats and wealthy landowners, Council officials, businessmen of all types and those from the south

'In Flight' magazine for the winter of 1984/5.

NEW AIRCRAFT, NEW ROUTES

(or European countries) who regularly fish, hunt deer, or shoot grouse in the Highlands.

This route had other little touches too. St Andrews Night, Burns' Night and other notable occasions were usually marked by a piper to play the passengers aboard, with the relevant dishes being served for the evening dinner - haggis, tatties & neaps and so on.

Mo played her part in arranging all these little 'extras' to keep the passengers interested, as well as her cabin staff. Constant innovation kept everyone on their toes! Getting up at 0430 hours in the middle of a Highland winter in order to be at the airport to crew the 0650 'red-eye' to London may be demanding, but Mo did it regularly, and was never bored by her work.

Apart from the efforts of the staff being appreciated locally, they have been noticed more widely in recent years. One well-travelled businessman from the USA wrote to say that he had never had a better breakfast on any other airline in the world; another, a resident of Wick (over 100 miles to the North) said he would rather drive to Inverness to catch the Dan-Air flight than take the early flight from Wick to Aberdeen and change there for Heathrow!

Proof of the pudding (or the beverage) came with the Tea Council awarding Dan-Air top prize for the best cup of tea in the world served - guess where - on the Inverness route!

Not every flight was routine. One of the Inverness team, stewardess Barbara Black, had good reason to remember one flight from Inverness to Manchester and to thank her safety and medical training. Just after take-off the heart of one of her passengers stopped beating and Barbara had to take swift action to revive him. For this she received a Royal Humane Society scroll from Inverness' Provost Alan Sellar and a gold watch from Dan-Air.

'In Flight' magazine for the summer of 1985.

Barbara Black receiving her Humane Society Scroll award from Provost Alan Sellar (left) and Chief Constable Hugh MacMillan of the Northern Constabulary. (DAS/GMS Collection)

137

DAN-AIR SERVICES

Grouse on the flight deck and Captain Ken Jones is not complaining!.
(DAS/GMS Collection)

Thus Mo Perera and her team became an integral part of the Highland scene, and the widely held adoption of Dan-Air as the Highlands' own airline owed everything to the way she and all the other Dan-Air employees at both ends of the route worked so unstintingly to keep up the quality of the service.

The Inverness - Heathrow route seems to have attracted more than its fair share of anecdotes since Dan-Air took over. Possibly one of the most interesting, demonstrates the lengths that the company's Reservations Department would go to in order to meet customer requirements...

On 14 February 1986, an RAF Squadron Leader flying Jaguar strike aircraft at RAF Lossiemouth tried by telephone to reserve a seat on DA157 INV-LHR the following evening for a meeting the next day but the Dan-Air flight was fully booked. The Squadron leader was desperate to get to London, and even tried ABZ-LGW, but that was also full, so he was put on the 'wait' list for possible cancellations. The next morning Dan-Air's Douglas Clark called Lossiemouth from Inverness to say that a seat was now available, and asked whether the Squadron Leader still required it.

At that moment Dan-Air's potential passenger was flying a training sortie somewhere in the wide blue yonder, so Lossie's ground controller called him on the radio, only to be asked for the Inverness radio frequency which Douglas provided. The next thing Dan-Air Inverness heard was a call on 130.65 mHz confirming the booking, direct from RAF Jaguar to Dan-Air!

'In Flight' magazine for the winter of 1985/6.

NEW AIRCRAFT, NEW ROUTES

Mrs Rita Purves of Culloden Moor (in light jacket) recives a champagne welcome aboard a company 1-11 and free tickets for her next flight as the 200,000th passenger on the Inverness to Heathrow service - just 30 months since Dan-Air took over the route! Also present are (L. to R.) INV Station Manager Gary Dominey, Area Sales Rep. Andrew Hamilton, A/S Kerry Reid, A/S Linda Dominey, Mrs Rita Purves, Martin Clough, Mo Perera and George Yeaman. On the steps are Mrs Purves daughters Shona & Eilidh. (DAS/GMS Collection)

"Good morning, Dan-Air..."

Whether a customer dialled 0345 100200 for the cost of a local call from anywhere within the United Kingdom and reached Reservations or, for some reason, rang 0293 820700, which was the airline's Headquarters at Newman House, Horley, each call was received efficiently and courteously.

Newman House, custom-built and opened on 28 January 1980 by Norman Tebbit MP, contained many of the administrative and technical offices required to run an airline. The Horley centre was divided into two divisions: the Customer Services Division, which handled such diverse topics as baggage claims, in-flight sales and catering, public relations, purchasing, route planning and ground services and the Sales and Marketing Division which handled international, IT, overseas, commercial and leisure sales, licencing, planning, ticketing, revenue management and industry affairs.

Sales and Reservations.

Reservations, an area made up of six departments, looked after both seat bookings and specific advance passenger requirements under the contol of Systems Reservations Manager Steve Brett.

Calls received by Telephone Sales from passengers wishing to travel in both directions on Dan-Air were easy to handle, but usually it was not as simple as that! Many wished to travel one way with Dan-Air, onward with another carrier or on another airline for the return journey. This was dealt with by the Fares Department of just four people, all fully conversant with international ticketing

DAN-AIR SERVICES

procedures, who provided fares quotations to Telephone Sales or the travel agents direct. The Fares Department was also responsible for ensuring that the airline's fares were accurately reflected in SHARES (the Dan-Air reservation system) and the SITA fare-share system, which provided fare quotation facilities for the majority of the UK travel agents via Galileo, the travel agents' Computerised Reservation System (CRS)

More passengers than ever required their tickets to be issued via the mail, and this was handled by the Ticketing Department. They issued over 10,000 tickets a week, and in return received over £600,000 a month in ticketed revenue. Ticketing also handled 'Eurocar' car hire arrangements, preferred hotel reservations and any special offers - like the *Daily Express* and *Daily Telegraph* 'two for one' promotion which ran very successfully in the summer of 1992.

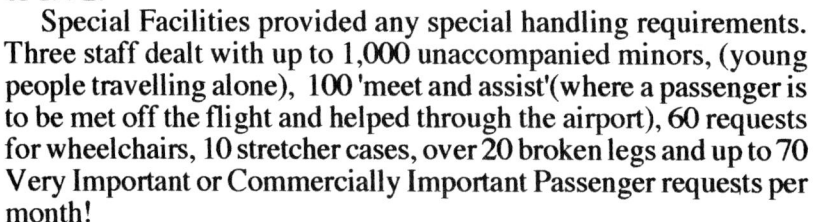

'In Flight' magazine for the spring of 1986

Special Facilities provided any special handling requirements. Three staff dealt with up to 1,000 unaccompanied minors, (young people travelling alone), 100 'meet and assist'(where a passenger is to be met off the flight and helped through the airport), 60 requests for wheelchairs, 10 stretcher cases, over 20 broken legs and up to 70 Very Important or Commercially Important Passenger requests per month!

Group Reservation Sales (GRS) handled any reservations of ten people or more travelling together. They had to be aware of their requirements and what other airlines offered and negotiated a fare which hopefully won each group for Dan-Air. Alongside the GRS was Contracts and Allocations who worked closely with UK Sales to ensure that the correct number of tour operators' seats were allocated to each flight.

What the travel agent saw when booking a flight on Dan-Air's CRS. A sample availability display, along with explanation.

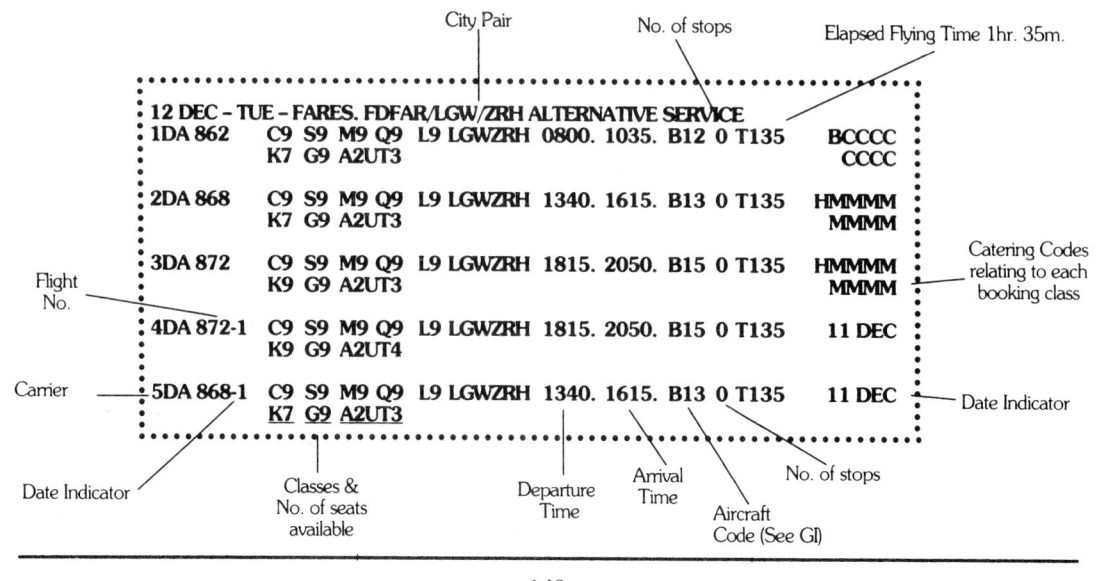

```
12 DEC - TUE - FARES. FDFAR/LGW/ZRH ALTERNATIVE SERVICE
1DA 862    C9 S9 M9 Q9  L9 LGWZRH  0800. 1035. B12 0 T135    BCCCC
           K7 G9 A2UT3                                        CCCC

2DA 868    C9 S9 M9 Q9  L9 LGWZRH  1340. 1615. B13 0 T135    HMMMM
           K7 G9 A2UT3                                        MMMM

3DA 872    C9 S9 M9 Q9  L9 LGWZRH  1815. 2050. B15 0 T135    HMMMM
           K9 G9 A2UT3                                        MMMM

4DA 872-1  C9 S9 M9 Q9  L9 LGWZRH  1815. 2050. B15 0 T135    11 DEC
           K9 G9 A2UT4

5DA 868-1  C9 S9 M9 Q9  L9 LGWZRH  1340. 1615. B13 0 T135    11 DEC
           K7 G9 A2UT3
```

Labels: City Pair, No. of stops, Elapsed Flying Time 1hr. 35m., Flight No., Carrier, Date Indicator, Classes & No. of seats available, Departure Time, Arrival Time, Aircraft Code (See GI), No. of stops, Catering Codes relating to each booking class, Date Indicator

NEW AIRCRAFT, NEW ROUTES

Reservations in Newman House: (clockwise from left) Linda Hudson, unknown, Nigel Parsons, Maggie Randall, Suzanne Hutchins, Frank Lefervre, Christine Watkins, Maureen Harrold, Sarah Mander, Nick Wooten, Jessica Fenn, Alision Bristow. (DAS/GMS Collection)

'In Flight' magazine for the summer of 1986.

Reservations System Control (RSC) conducted pre-flight checks to ensure that there were no duplicate bookings, all the names were in the bookings, and that all passengers had been ticketed. This was intended to prevent any problems arising at the check-in desks. RSC also checked that the aircraft configuration matched the potential load on the day. RSC were at the sharp end of any schedule change activity by re-routing or re-allocating the passengers on the next service available to minimise any inconvenience. Finally, they continually checked most of the world's CRSs to ensure that Dan-Air's flight schedules appeared and were displayed correctly.

Reservation Services provided support for all reservations offices world-wide. Accuracy here was essential if the Company was to maximise the potential of selling on a global scale. In 1992 70% of all Dan-Air's sales were made by a CRS, so it easy to see the importance!

Dan-Air had, in 1992, concluded an agreement with SABRE on of the world's largest CRSs - Direct Connect Air - that utilised state of the art EDIFACT technology for transmitting messages between systems. This placed the airline as the first in the UK to introduce this technology for subscribers of SABRE and, as a result, expected a 30% increase in booking through the use of this system alone!

Cabin Staff Training

As can be gathered from our look at the Inverness - Heathrow services earlier, the cabin staff had much more to do than just serve drinks and meals.

A member of Dan-Air's cabin staff was indeed a jack-of-all-

DAN-AIR SERVICES

trades - and highly skilled at many! Diplomat, lifeguard, firefighter, accountant, sales representative, caterer, psychiatrist, security guard, nurse, psychologist - all these titles have aspects contained in the cabin staff job description. They were perceived by passengers as servants, travel guides, fashion models...

Dan-Air was one of the first independent airlines to introduce a co-ordinated training scheme (with dedicated instructors and facilities) for its cabin staff. Because of the varied fleet operated by the airline cabin staff, members were trained (and therefore became qualified) to operate across the entire fleet, rather than on one or two specific types. This form of staff training was way ahead of what was required by the existing rules and regulations, which eventually caught up with the airline! Because equipment has became more and more complex, there was a certain amount of regression in recent years back to cabin staff being trained for one or two types, but nevertheless, the training was just as comprehensive and to the very highest of standards.

The selection of a potential member of the Dan-Air cabin staff team started with the application form. Previous experience in reception work, catering, or nursing was an advantage, as was the ability to speak one or more languages other than English, but none of these was essential. However good health (all successful applicants underwent a careful medial examination) and an education to certain standards was required.

Successful applicants were made up into groups and attended a day's interview at one of the Dan-Air base airports, where they met and talked informally to the Chief Stewardess and senior members of the Company's staff. The interviewers were looking for many things: of course maths, aptitude and general knowledge were tested, and perhaps sometimes too much was asked of the applicants, but it was better that they were sometimes wrong on the

'In Flight' magazine for the winter of 1986/7.

In the classroom! A course of cabin staff recruits undergoes training in the finer points of looking after passengers. (DAS/GMS Collection)

NEW AIRCRAFT, NEW ROUTES

6 March 1987 - a very important day for the Company, for that was when Cabin Crew Training Course Six completed four weeks' basic training. Nothing unusual about that perhaps, but it was the first course to include male stewards. The photograph shows the course members having just been awarded their 'Wings'.
(DAS/GMS Collection)

'In Flight' magazine for the spring of 1987.

ground than in the air later. After all, care of passengers came first, so the Company was looking for a balance of commonsense and initiative, the ability to recognise a problem - if possible before it arose - and then deal with it calmly and without fuss.

The initial course was a period in the classroom followed by a period of probationary flying and working alongside experienced cabin crews. Dan-Air's cabin staff were trained about their aircraft, the equipment in it and what sort of service they were to provide to the customer by the Cabin Services Department. Each intake was trained initially on two types of aircraft, according to the Company's requirements at the time.

As the main aircraft type in the Dan-Air fleet was slowly being standardised on the Boeing 737, let us look at what is involved in training a cabin staff member for just this type.

The initial course lasted three to four days. Following an introduction, the cabin facilites, toilets, galley and bar units, water system and vacuum sockets were explained. Then there was a sixteen point section relating to the location and use of emergency equipment - everything from lifejackets to seatbelt extensions, lifecots to megaphones. This was followed by the aircraft's emergency equipment - exits, lighting, ropes and slides. The aircraft facilities - cabin crew panels, galleys and audio warnings associated with the Call System were explained, as were the drills and procedures that covered everything from evacuation to the action needed to restrain an unruly passenger!

Crew seating positions, refueling with passengers aboard and cabin crew duties in the cabin were also explained, along with passenger briefings, flight deck checks and the seating of

DAN-AIR SERVICES

passengers. If a cabin staff member was transferring from another aircraft type in the fleet, the course still lasted three days and all members underwent a annual one-day refresher course.

Throughout the working life of a member of the cabin staff team, he or she underwent continual assessment, just as the flight deck crews. Each member of the cabin staff on every flight carried a number and specific duties were allocated to that number. The 'No.1', for example was directly responsible to the Captain for the efficient working of the cabin, and that covers just about everything from custody of the 'ships papers' to closing and opening the cabin doors.

A crew may have found one of their number was a 3-star Check Stewardess who would fly and work with them, or one of the Base Stewardesses responsible for cabin staff administration. Perhaps one of the staff on a flight was the Fleet Stewardess, responsible for all the work on a particular type of aircraft. There may have even been a crew member wearing the six stars of the Chief Air Stewardess. *'Those six stars'* as one stewardess remarked to the author *'concentrated the mind wonderfully!'*

'In Flight' magazine for the summer of 1987

More profit and more expansion.

During 1983 the parent Company announced an increase in profits of £3.3 million for the 1982 trading year, much of which had been brought about by the collapse of Laker Airways and the resulting reduction in competition. About 80% of all passengers carried were on charter services, the remainder going on the scheduled routes.

An announcement was also made during the year that Dan-Air was to apply for permission to develop a nine-acre site at Gatwick Airport for apron space, car parking, offices, a hangar and stores.

1984 saw more scheduled services start. From May, an aircraft was based in Jersey for a London Gatwick - Channel Isles link. On 14 May, a daily 1-11 Manchester - Zürich service began. By now the Group was recording an annual turnover of nearly £200 million and in that year carried 4.5 million passengers on a 53 aircraft fleet, about a quarter of which were on scheduled services. The boom time of the 1980s was beginning to make the future look good; it was time to strengthen the position even more.

DAN-AIR SERVICES

CHAPTER NINE
A new generation - a new service!

For the start of the 1986 holiday season the Company's largest aircraft arrived at Gatwick - an example of the Airbus Industrie A300 Airbus acquired from the German charter operator Hapag-Lloyd. Delivered at the end of April, the aircraft, which entered service with Dan-Air in May, could carry 312 passengers in the 2-4-2 twin aisle configuration and had a crew of three on the flight deck and nine cabin staff.

The A300 Airbus was the fourth of the 'wide-body' airliners to become operational, and the first to be produced in Europe. It was the first product of a truly international programme of development and production to reach commercial service.

The project was launched in 1965 as an Anglo-French initiative to develop a large capacity transport aircraft for BEA and Air France, at the same time as a German group of manufacturers was making a similar study for Lufthansa. The designs were combined in 1967 and evolved further into the final shape which took to the air for the first time in late 1972.

Dan-Air's first Airbus was registered G-BMNA, soon to be followed by a similar machine, also obtained from Hapag-Lloyd and registered G-BMNB. This example was an A300B4, which, through increased tankage, had a greater range, and it immediately entered service on Christmas holiday flights to the Mediterranean, Israel and the Canary Islands, configured with 336 seats.

Introduction of the new type was not without minor problems. The crew training programme meant that for the first six weeks or so of the 1986 summer season the airline would not have a full complement of Airbus crews. Consequently, it was not until the latter half of June that aircraft and crews matched the business sold. Nevertheless, the Airbus programme was particularly pleasing for, with the introduction of the 'wide bodies' it meant that Dan-Air

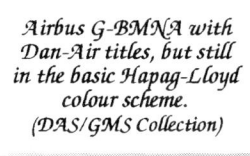

*Airbus G-BMNA with Dan-Air titles, but still in the basic Hapag-Lloyd colour scheme.
(DAS/GMS Collection)*

DAN-AIR SERVICES

A hold full of Hosties! The cabin crew pose in the rear cargo hold of an Airbus. This picture gives a good idea as to the size of the aircraft and its capacity! (Norma Franks)

could provide an even wider range of seating capacity to its clients. Now it could offer an aircraft with larger capacity for the peak times and destinations - Mahon on Friday, Palma on Saturday, Faro on Sunday, Corfu on Monday...

At last the airline had wide body capacity to pitch against British Airtours' Tristars, Calair's DC10s and Britannia's 767s. The type soon started turning in good figures, allowing the airline to claim the highest utilization per day for an Airbus operator anywhere; In June 1986 they averaged 11.93 hrs compared to 10.00 hrs, the next highest of the time. The staff that flew the Airbus also loved them; from the pilots viewpoint they were a dream to handle and the twin-aisle configuration made it so much easier for the cabin staff to serve the passengers.

Engineering back-up...

After every flight, no matter what the type of aircraft, pilots and cabin staff alike report any defects within their areas of responsibility that they have observed and logged on the flight; equally important, any suspected or possible defects were also logged. Everything was meticulously checked: engines, airframe, electrics, hydraulics and all other systems were inspected, faults rectified and double-checked before the aircraft was cleared for flight again. All reports, whether or not they could be dealt with locally, were fed back into the records maintained on each individual aircraft where they were noted for subsequent checks and inspections. The aircraft was then put through another series of pre-flight checks by the flight crew.

All this engineering work required a team of dedicated personnel, intimately familiar with all types of aircraft and their

A NEW GENERATION - A NEW SERVICE!

equipment that the airline was currently flying. Much of this 'line maintenance' was preventative, as it is always better to prevent than cure. This disciplined process of check, report, repair or replace, check and check again was the basis of Dan-Air's impressive maintenance record.

For every minute a Dan-Air aircraft spent in the air it was given some 2.5 minutes maintenance; in every year of its working life, an aircraft would have spent some 20 days undergoing routine the 'base maintenance' laid down for it.

From the moment of occupation at Lasham, Dan-Air Engineering set about to plan for gradual expansion, consolidating each stage fully before moving onto the next. A new hangar and office complex was completed there in 1970, the old buildings being bulldozed in preparation for a new No. 2 hangar. By 1990 the company had the use of three hangars with an area of 69,450 square feet, with provision for four aircraft bays, two of them capable of accommodating aircraft to the size of a Boeing 707. Lasham also housed the Engineering Division's training centre.

Not only were the Company's own aircraft looked after, so were machines belonging to many other airlines. Such contracted customers could draw on the services of experienced Dan-Air project engineers to obtain specialist advice and expertise on the full range of aircraft types covered by Dan-Air's Approval, granted not only by the British Civil Aviation Authority, but by the American Federal Aviation Authority and many others.

An extensive non-destructive testing (NDT) capability, approved by the Civil Aviation Authority, catered for the requirement of aircraft inspection at all three maintenance bases as well as crack detection work on a wide range of components. The NDT team utilised a wide range of X-Ray, ultrasonic, eddy current magnetic particle, or isotope techniques, both on and off-site.

The Manchester facility had been operating since 1976, when it

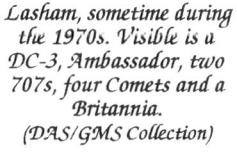

Lasham, sometime during the 1970s. Visible is a DC-3, Ambassador, two 707s, four Comets and a Britannia.
(DAS/GMS Collection)

DAN-AIR SERVICES

became clear that the airline's fleet had increased to a size with which Lasham would not be able to cope. Two hangars were used, capable of accommodating six short to medium range aircraft and were placed to good use for 1-11, HS 748, DHC-7 and MD80 series aircraft. The Dan-Air facilities at Manchester made it the largest airline employer after the Airport Authority. Dan-Air developed a whole range of approved maintenance schedules to which customers' aircraft could be maintained, including the planning and control from one maintenance schedule to another if required.

Computers were not only used on advanced flight-decks. At Lasham the computerised technical recording system was used to ensure that maintenance (both present and future) was recorded, with requirements properly planned and executed. The computer was also used for stock control over the extensive range of spares (either held or ordered) which kept Dan-Air and other airlines' fleets in the air. A 24 hour spares support organisation was constantly available to meet the 'aircraft on ground' situation.

In 1990 Dan-Air sold off its engineering side to the Danish FLS Aerospace company, but retained the new owners to maintain the fleet and look after the day-to-day servicing.

How did all this work in practice? Well, let us look at the various degrees of 'check' for a Dan-Air Boeing 727. Apart from the normal pre-flight check that the aircraft underwent from the flight crews, it also received either a 'Departure' or 'Arrival' check from the ground engineers; 23 items before departure and 16 on arrival. If an aircraft was parked overnight it was given an additional, slightly more detailed 'Terminal' line check (20 items) before its next flight.

Not more than every 32 days it was given a much more detailed 'Intermediate' line check in four parts, allowing the checks to be 'staggered' to avoid taking the aircraft out of revenue earning service for too long.

Then there were the 'Base' checks, which consisted of even more

The new hangar and maintenance facility under construction at Gatwick.
(DAS/GMS Collection)

A NEW GENERATION - A NEW SERVICE!

detailed procedures and certainly required the aircraft being taken out of use. A 'B' check was undertaken every 1000 flying hours or 186 days (whichever came first), a 'C' check every 14 months. The major 'D' check, where the 727 was virtually dismantled to component parts, inspected, repaired and re-assembled, this was done initially at 24,000 flying hours, then at 56,000 and 70,000 hours, the latter being 2,916 days, or eight years solid flying!

All the other aircraft types in Dan-Air's fleet underwent a similar sort of maintenance inspection procedure, with minor variations dependent on the type.

Of Advertising and Baggage...
1986 saw the launch of a nationwide television advertising campaign on behalf of the airline's scheduled services network. Called *'Going Places'*, the campaign featured a businessman extolling the virtues of the airline and was aired in the Tyne Tees area before moving to Granada, Thames, Ireland, Grampian and Ulster regions. The TV advertising was linked to other forms of media, along with taxi and bus-side promotion, which raised the profile of the Company and brought in much new business.

Also going places during 1986 was mountaineer Chris Bonington, who set the check-in and baggage staff at Newcastle a mountain of a task! Chris was setting off on an expedition to climb the as yet unconquered 23,565ft Menlungtse in Tibet.

He turned up on time at the check-in desk for the flight to Oslo with 700kg of baggage, potentially incurring a massive £1,700 bill in excess charges. Ever aware of the publicity value, Dan-Air waived the charges against the promise of a picture of the famous

A publicity photograph taken at Tower Bridge in London. The event was the launch of 'taxi-side' advertising on the city's black cabs.
(DAS/GMS Collection)

DAN-AIR SERVICES

Yeti, Menlungtse being where Eric Shipton took the famous 'Yeti-footprint' photograph in 1951!

Seen, but not heard...
Dan-Air also became the first operator of the British Aerospace (BAe) 146 regional airliner which, through its supremely quiet operation, demonstrated 'good neighbourliness', thus allowing the opening up of a whole new series of airports to jet airliners.

The honour of being the first airline in the world to place the new design into service went to Dan-Air when, as launch customer for the series -100, it took delivery of its first aircraft at a hand-over ceremony at Hatfield on 23 May 1983. After the ceremonial hand-over, during which the aircraft's log books were handed to the Dan Air Chairman, the Company's 146 Fleet Captain, Laurie Buist, flew the aircraft low over central London to demonstrate the low noise level which was so characteristic of the new design.

This was not the first time that the 146 had appeared in Dan-Air colours however, for the third aircraft built, number 1003 registered G-SSCH, appeared at the 1982 Farnborough Air Show to celebrate Dan-Air's order for two aircraft with options on two more. This machine had been used for many noise measurement trials and most of the performance testing. While in Dan-Air colours it also flew for a while with a long nose-mounted instrumentation probe in order to measure the undisturbed air ahead of the aircraft.

According to the airline's chairman, Mr F.Newman, speaking earlier in the year during a visit to British Aerospace's Civil Aircraft Division at Hatfield where the two aircraft on order were being built:

A beautiful picture of BAe 146 G-BKHT above the clouds. This photograph was to be used for publicity purposes, but, according to the scribble on the rear of the original, was rejected because the airline's name could not be seen clearly! (DAS/GMS Collection)

A NEW GENERATION - A NEW SERVICE!

G-SSCH, the evaluation aircraft in temporary Dan-Air colours at Jersey Airport. (DAS/GMS Collection)

"Dan-Air has chosen the BAe146 because of its unrivalled economy, efficiency and quietness... ...the wide cabin of the BAe146 offers a high standard of passenger comfort... ...and the performance of the aircraft will allow us great flexibility in using it on many of our current routes".

Prolonged design and development

The BAe146 suffered a particularly long and painful gestation period, with its roots based in the 1959 concept of a Dakota replacement by the then De Havilland Aircraft Company.

Conceived as a feasibility study numbered the DH123, the design was to be a high wing, twin-Gnome turboprop-powered machine carrying 32-40 passengers. However, public acceptance of the Comet and Boeing 707 convinced the designers at Hatfield that the future lay with the turbojet, so the design was re-worked during 1960 in favour of the DH126, a 30-seater with twin rear-mounted turbofans. In turn this developed into the HS131 after take-over into the Hawker Siddeley Group, making use of as many HS748 components as possible but retaining the basic configuration of the DH126.

This was succeeded by the HS136, a low-wing, rear engined T-tailed design with seating for around 40 passengers. Started in 1964, by 1967 the design had been stretched into a 50-60 seater with the rear engines abandoned in favour of podded, underwing

Mr F. Newman, Chairman of Dan-Air Services Ltd (seated right) and M Goldsmith, Divisional manager of BAe Hatfield-Chester, sign the contract for the first two 146 machines. Standing behind are Directors and Senior Executives from both Companies. (DAS/GMS Collection)

DAN-AIR SERVICES

Kim Denton serves refreshments to passengers aboard a Company 146-100. The wide-bodied look of the interior is particularly noticable (DAS/GMS Collection)

powerplants similar in design to the Boeing 737. Design changes occurred again, so that by 1969 the design designation was changed to the HS144, powered by a pair of rear-mounted Rolls Royce Trents, with a low-mounted wing and T-tail.

Finally, after problems with development of the Trent, the engines were changed yet again, this time making use of the Lycoming ALF502 turbofan mounted in pods under a high-mounted wing, but rampant inflation and rocketing oil prices forced an overnight slump in the airliner market and all work stopped, with only the smallest amount of government funds made available for continuance of design work.

For three years the project ticked over in low gear, during which time the majority of the UK's aircraft industry was nationalised under the banner of British Aerospace, bringing about a final change in designation - the BAe146. On 10 July 1978 the company announced full scale resumption of work as a private venture. Focal point for the manufacture was the old De Havilland works at Hatfield, but other BAe plants and other British and overseas suppliers would contribute to a high degree.

With growing complaints from the general public of excessive noise from jet aircraft, much care and thought went into making the aircraft as quiet as possible. So it was no surprise that when the prototype, aptly registered G-SSSH, took to the air for the first time on 3 September 1981 it was the quietest jet airliner ever built.

For the first time in a jet...
Ten years after beginning the Gatwick to Berne service, a route which many critics said would fail from the onset, Dan-Air had the last laugh when they placed the BAe 146 on the route. The aircraft was the only jet in the world capable of safely operating from Berne's 1,310 metre runway.

The 146 soon found favour with the Swiss; the airport authorities at Berne even went to the lengths of building a new

A NEW GENERATION - A NEW SERVICE!

departure building for 92 passengers which became necessary when the 146 service was introduced. The hall, built in July 1983, must hold a world record for speed of airport building construction, for it was completed in one and a half hours! The record time was achieved by the simple method of joining two large containers together and then removing the centre wall!

The many skiers that flew into the Alps every year also appreciated the BAe146, for in December 1984 two new destinations were added to the charter destinations served by the aircraft. Neither Innsbruck in the Tyrol or Chambery in the French Alps had ever been served by jet aircraft before. By as early as January 1985 Dan-Air recorded its one millionth BAe146 passenger and that with only three machines in the fleet!

Dan-Air placed the 146 on some of its turbo-prop routes as an experiment to see if pure jet aircraft could be justified. Initially, one aircraft was to be based at Gatwick for use on schedules to Berne, Zürich, Perpignan, Toulouse and Dublin. The second aircraft was to be Newcastle-based and operate to Gatwick, Bergen, and Stavanger - an important centre of the Norwegian oil support industry - during the week and also fly the Sunday evening schedule between Jersey and Leeds/Bradford.

The Newcastle-based aircraft was used on a 13 hour, 11 sector day, starting at 07.15hrs. Starting at Newcastle, the aircraft hopped over to Tees-Side, then Amsterdam, Bristol, Cardiff, the Channel Islands and finally Cardiff before backtracking down the route to Newcastle. The average block time was just 43 minutes, with an average time on ground of 15 minutes. The result was a dramatic increase in passengers, thus justifying the use of the type.

146 Mike November at Flughafen Innsbruck. The proximity (and height) of the mountains behind is somewhat noticeable! (DAS/GMS Collection)

DAN-AIR SERVICES

*146 Mike November taxies out for the start of another flight from Gatwick.
(DAS/GMS Collection)*

In 1992 full expansion into scheduled services brought about the decision to update the 146 and other types within the airline's fleet, to allow the aircraft to operate in Category II weather conditions. The two early production series 100 machines were returned to the manufacturer, and a number of series 300 all-weather machines were ordered. Despite Gatwick's importance as London's second airport, there are times when the weather can be quite foggy and thus put restrictions on aircraft movements or possibly force a diversion to another airport. Alan Lupton remembers one such flight where the inconvenience of a diversion actually worked in his favour:

> "I was coming back on a 146 from a business trip to Toulouse. As we crossed the coast we were told that, due to fog at Gatwick, we were to divert into London Heathrow. Although some of the passengers groaned, it didn't worry me too much. I live to the north of London anyway, and travelled to Gatwick by train, so I had no car to collect.
>
> A short while later we were told that Heathrow was too busy to take us and that instead we were going to land at Luton. There was an even bigger groan from those aboard at this titbit of information. I must have been the only one that thought, great! - almost door-to-door service courtesy of Dan-Air - I only live ten miles away from Luton!"

What is amusing about this unfortunate extended journey for most of the passengers on Alan's flight was that he worked at the time in 'Future Projects' of British Aerospace, so perhaps the experience had a hand in developing the new, all-weather variant!

By the time the 146 had been in service for six months it was possible to evaluate its performance much more accurately. Invariably there had been some snags and delays, but overall the type had performed well: nearly 3000 hours had been flown, made

A NEW GENERATION - A NEW SERVICE!

up of 2220 landings and an average of 6.6 flights per day. During the first month 12% of flights were delayed; during November that figure was down to 2.5%.

Operational economy was commercially vital, and in this respect the 146 was beginning to show its mettle - the type was burning 16% less fuel than the BAe 1-11 series 200! Modifications were carried out to improve passenger comfort and service even more, which enhanced an already popular aircraft.

By the time five years' experience had been gained, the first aircraft had completed 11,300 hours made up of 10,500 flights for the company. The fleet of three had made over 30,000 landings and had carried over 1.5 million passengers.

Promotion and Publicity...

Although Dan-Air has never been an airline to shout its own praise from the rooftops, that does not mean it has never pursued a policy of promoting the Company name to the public.

Throughout its history, Dan-Air's Publicity and Public Relations Department promoted the airline's name and logo; products such as a hot-air balloon, racing cars, even a world-class Scottish Pipe Band have all borne the Dan-Air name.

Possibly the highest public exposure the Company ever received occurred over the weekend of 13 - 15 July 1990 when the airline was heavily featured in '*Airport '90*' a Television South production for the ITV network which saw a series of live broadcasts from Gatwick Airport as part of the airport's 60th Anniversary celebrations. All aspects of the airport's operations were featured in

Captain John Hutchence and Captain David Reiss make their final checks before take-off from Newcastle on board this unidentified 146. (DAS/GMS Collection)

DAN-AIR SERVICES

the programmes, hosted by Nick Owen and Fern Britton, and Dan-Air allowed the television crews to 'adopt' and follow 'November Tango', one of their BAe 146 fleet, around Europe on its weekend's duties. In order to show what went on behind the scenes, they gave permission for the TV crews to enter the airlines most 'holy of holies' - the Operations Room.

A number of Dan-Air staff were interviewed or shown going about their jobs. For instance, Captains Vaughan Dow and Mike Stonehewer were shown introducing pilots to the 727. As the weekend sequence of programmes was broadcast during the busiest time of the year at the airport, one suspects that television crews were hoping for delays and problems to occur, so that they could show 'high drama' of the 'backroom boys' frantically trying to hold everything together whilst bored, tired passengers listlessly hung around the airport waiting for a flight. That may have made good television, but sadly for the televison crews, at the end of the weekend 'NT calmy arrived back 'on chocks' at Gatwick within minutes of its scheduled time having flown over 10,000 miles, carried nearly 3,000 passengers and maintained its timing across the entire weekend. There was no fuss, no drama, just a professional company doing its job!

The change of emphasis to that of scheduled operator brought the need to make both the public and the travel industry aware of the many new services and destinations, so a whole series of advertisements in the national and international Press was run.

Larger, quieter 'Baby Boeings'

Such was the success of the Boeing 737 that many airlines required it to have an even larger passenger-carrying capability. Nevertheless, there was growing awareness of noise pollution and although the whine of the Pratt & Whitney (P&W) JT8s had been

A flight of fantasy! Anne Murray (left) and Kate Young from the Aberdeen School Orchestra serenade Dan-Air senior ticket officer Linda Halligan on the flight-deck of this 1-11 at Aberdeen Airport. Simple publicity of this nature (the orchestra was flying Dan-Air) could be easily used to put the airline in the public eye. (DAS/GMS Collection)

A NEW GENERATION - A NEW SERVICE!

reduced by the 'Quiet Nacelles' there were still problems. Increasing the aircraft size demanded more power, so a major re-design was undertaken, resulting in the Boeing 737-300 series.

The visible appearance of the aircraft changed as the fuselage was lengthened by 104 inches over the -200 model. The powerplants were changed, with the slim nacelles housing a pair of P&W JT8s giving way to the more bulky flat bottomed nacelles containing CFM 56s developed by CFM International, a company jointly operated by General Electric in the USA and SNECMA in France. Use of these engines gave an increase in efficiency, reducing the seat cost per mile by 21%. The tail shape changed also; gone was the graceful, slightly curved leading edge of the fin, replaced by an angular leading edge. Other technological changes were made, including a computerised flight-management system, lightweight materials in the wing (with an improved aerodynamic leading edge), control surfaces and interior.

In June 1986, Boeing announced yet another version - the 400 series - with a 114 inch greater fuselage length than the -300 machines, but there were deeper, more important changes. With two-crew operation becoming the norm, flight crew workload had been increasing for some time. To reduce this, a much greater emphasis was placed on electronic flight deck instrumentation - the so called 'glass' cockpit. Many of the conventional dials and guages were replaced and amalgamated into a number of switchable, multi-purpose cathode ray tubes (CRTs).

The -300 series first flew in February 1984, the -400 series taking to the air for the first time on 19 February 1989. Dan-Air began to operate a mixed fleet of -200, -300 and -400 series aircraft. A good idea of the comparative sizes of these different versions can be gained by looking at the relative maximum take-off weights of each type. The -200ADV is approaching 50,000 kgs; the -300 series nearly 57,000 kgs and the -400 series between 64,500 and just over 68,000 kgs.

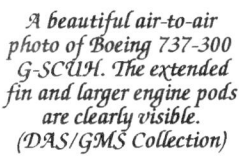

A beautiful air-to-air photo of Boeing 737-300 G-SCUH. The extended fin and larger engine pods are clearly visible. (DAS/GMS Collection)

DAN-AIR SERVICES

G-BNNK, a 737-400 outside the newly completed hangar and maintenance complex at Gatwick.
(DAS/GMS Collection)

The first Boeing 737-300 for Dan-Air was delivered in May for charter work, primarily in order to fulfil the requirements of travel agent Thomas Cook, for Dan-Air was now handling the bulk of their short-haul business.

Arrival of the first -400 series occurred during the murky evening of 1 December 1988 when Captains Malcolm De Garis and Simon Searle with fleet stewardess Val Barnett arrived from Seattle after night-stopping at Gander in Newfoundland. Also aboard were representatives of the airline, the Managing Directors of Owners Abroad and Redwing Holidays and the Aviation Director of Intasun. Six days later the aircraft entered service, flying to Gerona for Saga Holidays.

Forward Planning and Operations...
One of the most important centres within the Company on a day-to-day basis was the Dan-Air Operations Control Centre (DANOPS). It was from a small complex of offices, located on the third floor of Concorde House adjacent to the South Passenger Terminal at Gatwick, that the entire operation of the airline's aircraft and crew movements was conducted. This was done in a manner similar to a world-class conductor leading a symphony orchestra - in the airline's case the 'conductor' being the Duty Manager - Operations, ably assisted by the rest of the on-duty team whose tasks were to keep the programme together!

Here, minute by minute, every movement of every aircraft in the airline's fleet was planned, scheduled, monitored and recorded, both on a high technology computer system named SCOPE, that had been specially developed and designed totally in-house by the Company's management, with assistance from the Data Processing Section. Despite the high-technology, DANOPS also still used the tried and tested manual aircraft movement boards.

Aircraft movements were worked out months in advance by a

A NEW GENERATION - A NEW SERVICE!

number of Forward Planning Sections in Newman House. Movements of aircraft participating in the Company's scheduled services were reasonably regimented, because, by their very nature, they occured with a degree of regularity.

Charter planning was conducted in a similar manner, although requirements might not be the same year on year. What many passengers did not realise was that planning for their summer flight to the sun began, as far as the airline was concerned, about a year before that particular holiday season started!

The first indication the travelling public have of a forthcoming summer holiday season is when the tour companies brochures are placed on display in their local travel agents, usually in September. However, Dan-Air had already been talking to their main charterers since the previous June. As these discussions proceeded, the Company started to build up the next summer's charter programme, at this stage very roughly. The next important event happened in November, when all the airlines got together at a conference to co-ordinate the airport take-off and landing slots for the following summer at the busiest airports. Once these slots had been agreed Air Traffic slots - small periods of time allocated to allow an aircraft to take flight - had to be obtained to coincide with the airport slots, but this was an 'on the day' task.

The problems over slots became much more complicated in latter years. Most of the European states controlling routes over their countries exercise stricter controls over aircraft movements; at one time the only requirement was for the pilot to request permission to start up and, once granted, he could go. The negotiations for airport slots takes a while to resolve, but slowly over the winter months the programme come together as the airlines compromise on their requirements.

DANOPS.
The daily aircraft status boards are on three of the four walls.
Ops Controllers sit at terminals controlling it all...
(DAS/GMS Collection)

DAN-AIR SERVICES

Scheduled periods of aircraft maintenance, when they are taken off the 'line' are also known about well in advance and are normally timed to occur during the quieter winter months.

Parallel to scheduling Dan-Air's aircraft needs, the establishment planning of flight deck and cabin crew requirements occurred, taking into account licensing requirements, differences in aircraft types and a multitude of other mandatory requirements. Initially, provisional crew rosters were planned so that the number of aircrew required for each fleet could be worked out, but about eight weeks before the flight was scheduled, names were added to the lists, with rosters sent out to individuals. Crewing in Operations Control took control of these rosters about 10 to 14 days before take-off and from that point on had the responsibility for all other changes and 'fine tuning' until the flight was completed. This could be a mammoth task on the days when there were problems, when crews' needed to be held at home, changed when not 'in hours' for their next flight, or found for extra requirements.

The Crewing Controller was responsible for this area, with individual Crewing Officers working specific fleet establishments and a Hotac Section arranging hotels and transport at the airports involved.

Come the day of the flight, all the pertinent details appeared on both the computer system and the manual aircraft movement boards displayed on three walls of the Ops Room. All times shown on both systems were in Universal Time Standard (UTS) with data displays on the boards colour coded, dependent on the type of information - in red or green the data was factual, whilst blue was an estimate or forecast.

Information to update these boards - which occurred 'as it happened' - came in from a number of different sources - telephone, telex and radio via the Company's own VHF frequency. This allowed instant access to the current 'state of play' of all the aircraft, what they were doing and where they were. Often a potential problem in aircraft movements could be spotted by the Operations staff before it grew to major proportions; nevertheless, major problems could and did occur. If this happened during the peak season, just one machine suffering a relatively minor delay in getting away could generate a 'knock on' effect that reverberated around the airline's planned schedules for hours, until Ops hatched

Aircraft 07 - B.737-400

DATE	TIME	A/F	FLT.NO.	A/F	TIME
	0545	BRU	DA411	LGW	0645
FRI	0830	LGW	DA985	CDG	0930
28	1025	CDG	DA988	LGW	1125
AUG	1230	LGW	DA989	CDG	1330
	1415	CDG	DA992	LGW	1515
	1630	LGW	DA995	CDG	1730
	1815	CDG	DA998	LGW	1915
	2245	LGW	DA4142	GRO	0035
	0135	GRO	DA4143	LGW	0335
SAT	0830	LGW	DA985	CDG	0930
29	1025	CDG	DA988	LGW	1125
AUG	1230	LGW	DA989	CDG	1330
	1415	CDG	DA922	LGW	1515
	1630	LGW	DA995	CDG	1730
	1815	CDG	DA998	LGW	1915
	2130	LGW	DA4360	TFS	0145
	0245	TFS	DA4361	LGW	0700
SUN	0900	LGW	DA512	ATH	1235
30	1345	ATH	DA513	LGW	1730
AUG	1850	LGW	DA999	CDG	1950
	0555	CDG	DA980	LGW	0700
MON	0830	LGW	DA985	CDG	0930
31	1025	CDG	DA988	LGW	1125
AUG	1230	LGW	DA989	CDG	1330
	1415	CDG	DA992	LGW	1515
	1630	LGW	DA995	CDG	1730
	1815	CDG	DA998	LGW	1915
	2220	LGW	DA2738	PMI	0040
	0135	PMI	DA2739	LGW	0350
	0830	LGW	DA985	CDG	0930
TUES	1025	CDG	DA988	LGW	1125
01	1230	LGW	DA989	CDG	1130
SEPT	1415	CDG	DA992	LGW	1515
	1605	LGW	DA524	FCO	1830
	1920	FCO	DA525	LGW	2150
	2250	LGW	Maintenance		
		LGW	Maintenance		0520
WED	0645	LGW	DA790	AMS	0750
02	0835	AMS	DA793	LGW	0945
SEPT	1140	LGW	DA683	MAD	1350
	1455	MAD	DA684	LGW	1705
	2030	LGW	DA2302	AGP	2305
	0005	AGP	DA2303	LGW	0240
	0830	LGW	DA985	CDG	0930
THURS	1025	CDG	DA988	LGW	1125
03	1230	LGW	DA989	CDG	1330
SEPT	1415	CDG	DA992	LGW	1515
	1630	LGW	DA995	CDG	1730
	1815	CDG	DA998	LGW	1915
	2000	LGW	Maintenance		

A typical 'aircraft line' during 1992 for aircraft number 07. The three digit flight numbers denote scheduled services, while four digit numbers show the charter services (which in this case are all operated overnight.).

A NEW GENERATION - A NEW SERVICE!

Aircraft No. 03 - B.737-400

DATE	TIME	A/F	FLT.NO.	A/F	TIME
FRI 28 AUG	2335	LPA	DA2139	NCL	0405
	0630	NCL	DA2158	MAH	0915
	1005	MAH	DA2159	NCL	1255
	1425	NCL	DA2144	MLA	1755
	1850	MLA	DA2145	NCL	2230
	2330	NCL	DA2136	PMI	0210
SAT 29 AUG	0300	PMI	DA2137	NCL	0550
	0700	NCL	DA2124	PMI	0925
	1010	PMI	DA2125	NCL	1310
	1440	NCL	DA2148	ALC	1730
	1820	ALC	DA2149	NCL	2110
	2225	NCL	DA2152	CFU	0145
SUN 30 AUG	0235	CFU	DA2153	NCL	0610
	0715	NCL	DA2186	AGP	1015
	1110	AGP	DA2187	NCL	1415
	1530	NCL	DA2172	FAO	1835
	1925	FAO	DA2173	NCL	2225
	2340	NCL	DA2128	IBZ	0225
MON 31 AUG	0315	IBZ	DA2129	NCL	0605
	0720	NCL	DA2146	ADB	1130
	1120	ADB	DA2147	NCL	1640
	1740	NCL	DA2176	LCA	2225
	2315	LCA	DA2177	NCL	0425
TUES 01 SEPT	0710	NCL	DA2170	HER	1120
	1210	HER	DA2171	NCL	1640
	1755	NCL	DA2102	TFS	2240
	2330	TFS	DA2103	NCL	0410
WED 02 SEPT	0600	NCL	DA2116	RHO	1020
	1110	RHO	DA2117	NCL	1545
	1700	NCL	DA2134	PMI	1930
	2030	PMI	DA2135	NCL	2320
	2340	NCL	Maintenance		
THURS 03 SEPT		NCL	Maintenance		0515
	0615	NCL	DA2142	ACE	1040
	1130	ACE	DA2143	NCL	1555
	1755	NCL	DA2138	IPA	2235
	2335	LPA	DA2139	NCL	0405

The same 'aircraft line' for '03' (G-BSNW) operating out of Newcastle (NCL) on charter flights.

a plan to bring that particular aircraft line back on time!

Within the Ops Room, the Operations Controller helped the Duty Manager - Operations in his task, with other junior staff in assistance. The Customer Services Officer liaised with the charterers/tour operators to ensure they were aware of any problems on their flights, organised passenger welfare during delays and generally collated passenger figures and special requirements.

The staff in DANOPS worked twelve hour shifts, fours days on, four days off and the work could be extremely intensive as all problems tended to happen at the same time. Michael Forsyth, Dan-Air's Operations Manager, neatly summed up the section's task:

> "In an ideal world, we would just sit back and watch it all happen. Unfortunately, this seldom happened and problems directly out of our control conspired to cause disruptions, the biggest by far being Air Traffic Control delays. With upwards of 200 slots on a busy day to get, this could be a time consuming process. Other problems in this category could be industrial action or weather related. More controllable were Company contributable delays like technical defects on aircraft and crewing problems, however caused.
>
> The end result was that it could be a vast juggling act as aircraft type changes and tail numbers are altered to ensure that the programme operated as punctually as possible whatever the problems may have been. Its was a job where you had to think on your feet; there was no book of rules as each situation would have slightly different parameters; our senior staff had at least ten years experience. Something that made the job so interesting was that no one day was the same as the next".

The pilots and the operations staff of Dan-Air have always been regarded in the industry as being highly professional, but for the last twenty five years or so one phrase has greatly puzzled and annoyed many outside the company. Now, at last, the explanation can be given! Many pilots have been puzzled over the years about the use of 'Chug-a-lug' as either a sign-on, sign-off, or sometimes in the middle of an RT transmission. Bob Willis (former Dan-Air Operations Manager) explains what it was all about...

> " Many non-Dan-Air pilots have pondered its meaning over the years. Slowly, as the phrase became more commonly used, much suspicion built up that Dan-Air crews had this 'arrangement' with ATC. Many outsiders swore that chug-a-lug was a secret code-word requesting a direct track, or a better altitude. Others thought that its use brought Dan-Air a priority service.
>
> Well, now the truth can be told. Whilst the actual word 'chug-a-lug' is quite old, its use in Dan-Air stemmed from the post-British Eagle days when Dan-Air obtained its first 1-11s and based them at Luton. The phrase came into being when a few of us started to use it as a greeting in place of Good Morning or Good Evening. Later, it was used in the bar, as in Cheers, or Good Health.
>
> Slowly it spread around the airline's network as the staff moved about. It was also used during the numerous visits by perations staff and pilots to overseas as well as UK Air Traffic Control units. So 'Chug-a-lug' became a catchword used during

DAN-AIR SERVICES

the social parts of the visits; in the pubs, bistros, tavernas and auberges, and through these meetings it became used over the airwaves, in particular with the Eurocontrol Guild of Air Traffic Controllers at Maastricht. So, despite what others thought, 'Chug-a-lug' was no more than a cheerful greeting or sign-off!"

Aircrew Training

"A very high standard of training underpins the whole operation of an airline, and our training, for both Flight Deck and Cabin Staff is second-to-none." So spoke Dan-Air's Captain Jim Adams. Deputy Chief Pilot - Training.

Most of Dan-Air's training was conducted in-house, much at a departmental level. The airline was justifiably proud of its policy of taking on board responsibility for crew training; even at a very early stage of the introduction of a new type. In this way the Company could directly control the training to ensure that the high standards demanded were met. So good was the result that Dan-Air not only trained its own crews, but use of its facilities, training staff and methods was highly sought after by many of the world's airlines.

Dan-Air's training programme was divided into two dedicated areas, flight deck and cabin staff, each member being thoroughly and completely educated in their particular area of involvement. Vic Blake sums up the attitude towards training:

"From the moment a pupil comes into contact with the training staff, the whole thing is a friendly experience - there is no 'trapping' in it - it's intended to help the weakest one to achieve the airline's standard required. This is helped by the fact that everyone is here because they passionately want to do the job - we have no trouble at all with motivation!"

Most courses contained between ten and fifteen pupils. Flight crews were not trained to fly on these courses - they already knew how to do that; instead they received intensive, detailed instruction on how to operate the particular aircraft type they were about to be endorsed to fly. The detail and timescale varied with their own experience and the new type, but as an example, what follows related to the Boeing 737-400 conversion course.

Initially, pupils had twelve days in the classroom - referred to as 'Technical Ground School' - to teach the basic aircraft systems. This was followed by five hours on the Flight Training Device (FTD) to gain hands-on experience of where everything was located and how it operated; then another day in the Ground School. The student then sat the CAA Technical Exam at Aviation House, Gatwick. Next day the pupil made a 'supernumary flight' as a passenger in the flight deck of an actual flight to see the aircraft in operation. A further two days were then conducted in the FTD and a full day's safety training and ten days rostered in a full simulator. After this the pilot received his licence endorsement - 29 days of intensive learning and practice in all.

Bruce Fardell, one of Dan-Air's instructors, explains a new aspect to flight-deck crew training:

A NEW GENERATION - A NEW SERVICE!

"1992 was the first year that Dan-Air's more experienced crews were converted to a new type using what is termed 'Zero Flight Time'. This was where pilots were taken through the technical school stages and onto a simulator via the FTD where the remainder of the conversion course took place.

This meant that the first time a newly converted-to-type pilot was flying a 'real' aircraft there were paying passengers on board. This type of cost-efficient training, many times lower in cost-per-hour terms than using an actual aircraft, was only possible by the incredible sophistication of today's flight deck simulators. Even the phrase 'Virtual Reality' is redundant in this context; this is 'Total Reality'. Dan-Air's flight simulators provide flight deck crews with an environment that is real and one where emergencies can be experienced and practiced time and again in perfect safety.

The simulators are also used for LOFT exercises - Line Orientated Flying Training - where a crew can can fly in real time from say Gatwick to Paris and experience all the visual, intrumentational and communication references over the entire 'journey' as if they had actually made the trip. The instructor can feed in any number of emergencies at any stage, and also 'freeze' the flight to explain additional nuances of the training, then 'reverse' the flight to allow the pilots to do that aspect again. Pilots are able to do much more in the simulator than ever would be possible in a real aircraft and in which they would certainly not have the facility to 're-do' a particular point!"

The first two flight deck staff to complete the ZFT course were Captain Mike Potts and Senior First Officer Richard Fenn, both with around 10 years experience on the 727. After time in the Training School they then went up to Coalville near Derby to spend about 44 hours each in British Midland's 737-400 simulator.

Both 'departments' aboard an airliner then came together for what was probably the most important aspect of their training schedule - Flight Safety. By training both teams at once, commonality and continuity could be achieved to ensure that all members of an aircraft's crew 'spoke the same language' and fully understood each others problems no matter what occurred.

Part of this involved the use of Dan-Air's Cabin Escape Trainer - fitted out as a small section of an airliner cabin, with seats, galley area (which also contained a smoke generator), main door, over-wing exits and an Escape Slide so that crews could experience first hand and practice what it would be like to evacuate the cabin.

Other flight safety training equipment included an area for practice of the recognition and extinguishing of different types of fires (using real fires) in the cabin using the correct type of fire extinguishers. Ditching drill, required in the event of an aircraft coming down in water, was fully explained and practiced in Horsham swimming pool!

John Craig explains another aspect of Flight Safety Training; the 'black museum'

"This was an area that contained an amazing display of actual examples of 'prohibited items to be taken aboard an aircraft' that had been confiscated from passengers and was used to brief cabin staff as to what items some passengers tried to take on board. These include fake and real guns, knives of all lengths, lighter fuel (both cans of petrol and aerosol butane) nail varnish remover, books of matches that have ignited - even five litre drums of yacht varnish!"

DAN-AIR SERVICES

DAN-AIR SERVICES LTD	**TRAINING MANUAL**

PILOT CONVERSION TRAINING RECORD

RANK/NAME .... A/C TYPE ....
LICENCE TYPE AND NUMBER .... COMPUTER NUMBER ....

COURSE PHASE	INSTRUCTORS* SIGNATURE
GROUND TECHNICAL COMMENCED __/__/__ COMPLETED __/__/__	
CAA EXAM PART I ARRANGED FOR __/__/__ RESULT _____ PART II ARRANGED FOR __/__/__ RESULT _____ ARRANGED FOR __/__/__ RESULT _____	
FLT. PLANNING/PERFORMANCE COMPLETED __/__/__	
MNPS OPERATIONS/ONS/INS/FMS* COMPLETED __/__/__	
LOADING COMPLETED __/__/__	
ROUTE AND AIRFIELD COMPLETED __/__/__	
FLIGHT SAFETY AND SURVIVAL COMPLETED __/__/__	
AVIATION SECURITY/DANGEROUS GOODS COMPLETED __/__/__	
C.P.T./SIMULATOR TRAINING COMMENCED __/__/__ COMPLETED __/__/__	
BASE TRAINING COMMENCED __/__/__ COMPLETED __/__/__	
CAA FORMS 1179 AND 1179A COMPLETED __/__/__	
COMBINED IRR/COMPETENCY TEST COMPLETED __/__/__	
INITIAL LINE CHECK COMPLETED __/__/__	
TYPE ENDORSED IN PART I DATED __/__/__	
LINE FLYING COMMENCED __/__/__ COMPLETED __/__/__	
INTERMEDIATE LINE CHECK (CAPTS ONLY) COMPLETED __/__/__	
FINAL LINE CHECK COMPLETED __/__/__	
FLIGHT COSTS (CAPTAINS ONLY) COMPLETED __/__/__	
INTERVIEW WITH CHIEF PILOT (CAPTAINS ONLY) COMPLETED __/__/__	

A signature in the right hand column indicates that the trainee has been trained and testsed in accordance with the Company Training Manual.

I CERTIFY THAT HAVING EXAMINED THE TRAINING RECORDS TEST FORMS AND CERTIFICATES RELATING TO. .... I AM SATISFIED THAT HE IS COMPETENT TO CARRY OUT THE DUTIES OF.............................. ON THE........................ AIRCRAFT.

DATE __/__/__   SIGNATURE................................DEPUTY CHIEF PILOT - TRAINING

ACCEPTED FOR LINE SERVICE...FLEET MANAGER

OPERATIONAL CLEARANCE CERTIFICATE ISSUED __/__/__
*Delete as applicable.

A NEW GENERATION - A NEW SERVICE!

To give an idea of how thorough Dan-Air's training schedule was, left is the pilots conversion training record card, to be completed in 29 days.

'Blue Riband' routes...

Following the merger of British Caledonian Airways into British Airways in March 1988, Dan-Air revealed that it was applying for some of the former's old routes, two of which were international.

By adding Paris (Charles de Gaulle) into the network the airline would serve five international capitals from Gatwick, the others being Berne, Dublin, Lisbon and Madrid. Services to Nice, with potential for both business and holiday traffic, would strengthen the other French destinations.

Following the granting of an unrestricted licence to Paris, the Board decided that the new route warranted a new level of service. The June 1988 edition of 'Flightline' revealed details under the banner 'Autumn in Paris', although it was obvious that the Company was still holding the fine details of what was planned close to its chest. "Flightline' reported that:

> "To ensure our competitors are not alerted to details of our product some of the more interesting aspects will not be detailed until nearer the launch date; however, we plan to
> - operate six roundtrip services a day, seven days a week.
> - introduce a Business Class cabin with its own brand name.
> - introduce new customer service standards.
> - introduce enhanced ground products for Business Class passengers.
> - support the launch with public relations and advertising campaigns both in the UK and France."

The piece went on to state that the battle from rivals on the Paris route was just beginning, and the stakes were high. During the first year, the Company intended to carry 250,000 passengers.

The inauguration of the Paris service took place on 2 October when a 1-11 was seen away by members of the Board. The occasion was also the first public outing by the Dan-Air Pipe Band.

...and the Class 'Elite'

With the shift in market emphasis and growing importance of the scheduled service network and its use by business travellers, a demand arose for a passenger class other than standard. Designed specifically for the executive traveller, 'Class Elite' - first introduced on the Paris and Nice routes - quickly became established as a European business class service that offered an outstandingly high quality of exclusive services and facilities.

The summer 1986 timetable (30 March - 25 Oct)

On the ground dedicated check-in facilities and executive lounges were opened at most airports served, along with a priority baggage reclaim. On the aircraft 'Class Elite' cabins were designed to give passengers maximum room and comfort with the fitment of special Space Generator seats to create a spacious 2 - 2 configuration, with the middle seat - guaranteed unsold - converting to a wide table. Complimentary champagne or orange juice greeted the 'Class Elite' travellers, who could enjoy specially prepared meals with menus created by leading chefs.

DAN-AIR SERVICES

John Varrier (left) and Vic Sheppard try out the mock-up of the new 'Class Elite' seating, whilst Kay Grewal provides the refreshment. (DAS/GMS Collection)

Obviously the creation of two classes of seating brought about changes to the number of passengers that could be carried; on the BAC 1-11s 'Class Elite' passengers could be between 16 and 48, with between 39 and 79 in economy, depending on the aircraft used. The 146-300 had between 16 and 40 'Class Elite' with 46 to 82 Economy. The 737-300 was equipped for between 16 and 48 'Class Elite' and 66 to 114 Economy, whilst the -400 model had 16 to 52 'Class Elite' and 78 to 132 Economy.

The variation of cabin layouts across the fleet brought tremendous flexibility: if a particular route was known for the carriage of a high number of 'Class Elite' passengers, then an aircraft with an appropriate number of seats would be used.

The importance of the 'Class Elite' service to the Company's financial success was succinctly explained by David James at the Annual General Meeting on Friday 29 May 1992.

> "...movement of £20 million in profit or loss would be the equivalent to the revenue resulting from Dan-Air carrying, say, just two extra or two fewer passengers in the Elite Business Class section every time a plane takes off throughout the entire year".

Awards and recognition...

Close study of the 1985 Reports and Accounts presented to the shareholders revealed one interesting little snippet of information, the Chairman, Mr F. E. F. Newman had been Invested as a Commander of the Order of the British Empire (CBE).

But the Chairman was not the only member of the airline to receive an award. Although no longer part of Dan-Air, the news that Allan Snudden had also been invested with the CBE was

A NEW GENERATION - A NEW SERVICE!

*Left to right: Vic Sheppard, John Varrier, Sir Ian Pedder and Duty Manager John Fraser pose for photographs during the opening launch of 'Class Elite' check-in facilities.
(DAS/GMS Collection)*

received with much happiness. In 1991 Peter Somers, Associate Director of Planning was invested into the Order of the British Empire (OBE).

Caring for the customers...
Someone once wrote that the most complex journey any human can ever make is to travel from point A to point B by air. Many passengers understandably give no thought to the intricate route they are guided through, nor realise the diverse nature of the infrastructure that keeps a modern airline in business.

Any traveller's view of an airline tends to be naturally focused towards one particular aspect - the aeroplanes and the people who operate them. Although understandable, this is a very narrow viewpoint, for it is better to compare any airline with an iceberg - only about ten percent is visible, the rest is under the surface!

This was certainly true for Dan-Air. The public persona of the airline may have been of conservatively uniformed pilots, chic cabin crews and efficient ground staff, but that was just the visible pinnacle of the operation; behind the scenes there existed a huge, highly professional organisation, dedicated to ensuring that every aspect ran as smoothly as possible twenty four hours a day, seven days a week. Whilst the analogy to an iceberg is correct, it is also fair, and probably more accurate in many respects, to compare an airline at times to a swan - graceful, calm and serene above, whilst

DAN-AIR SERVICES

Over the years a number of in-house newspapers kept the staff informed. These ranged from the photo-copied 'Compass' (left) from March 1981 to the quarterly, then monthly "Flightline" (June 1988) that, in its latter guise contained many full colour photographs.

paddling like anything to keep going below the surface!

This support structure, although disregarded by most, is vitally important for the safety and smooth progress of the travelling public, and to provide a 'safety net' that quickly swings into action when a delay or problem occurs. This is especially so when caring for passengers on the ground.

Every airport to which Dan-Air flew had a number of people who were tasked to look after and smooth the passage of passengers while they were in the airport terminals. Some were empoyed by the airline, others worked for a handling agent contracted to provide this service to the Company. To demonstrate the wide variety of skills needed and the tasks that had to be undertaken, let us look at Dan-Air's ground operation at their busiest airport - Gatwick.

Nowadays all airlines provide a broadly similar product to their customers. All have to provide and maintain a reasonable schedule, reasonable check-in facilities and reasonable in flight service. What motivates a customer to fly with one airline again and again is how that airline provides services over and above what would normally be expected; how the inevitable delay is handled; how queues are minimisde; the friendliness of the service. Although Dan-Air's staff always tried to provide travellers with touches of personal service, the advent of emphasis on scheduled services in recent years causedthis aspect to become even more important.

The popularly used generic term 'ground staff', or, more accurately 'Passenger Service Representatives' (PSRs,) covered a whole host of different duties - staffing the check-in desks, boarding gates, scheduled services on-airport tickets sales, answering queries, solving problems, looking after 'Class Elite'

A NEW GENERATION - A NEW SERVICE!

Not the result of too much in-flight hospitality, just twins Jane and Sue Blackburn taking care of passenger John Lockwood. (DAS/GMS Collection)

passengers in their own lounge, carrying out the myriad of administrative duties. PSRs had to be able to cope with long hours of duty: twelve hours at a shift was common and required a person with just the right sort of attitude and outlook on life!

The staff at Gatwick never knew what duties they would tackle when reporting for a shift. The team, led by a Duty Manager, consisted of a number of PSRs who were dedicated to providing customer service for all the varying needs and demands of Dan-Air's diverse passengers.

On a peak summer day the staff would co-ordinate up to 70 flights, which amounted to a lot of passengers to deal with when there was a hiccup in the plan. One of the worse nightmares for airport staff was when the weather closed the airport; this would mean aircraft diverted to other airfields, passengers coached

DAN-AIR SERVICES

around the countryside likely to miss connections, and relatives frantic for information. Allison Beedie, Dan-Air's Airport and Passenger Service Manager...

"It is at times like these that the 'Spirit of Dan-Air' emerged. I remember many occasions when staff would work for an entire shift without even a cup of tea and never once did any of them complain. Staff would be despatched to hotels with young flyers or elderly travellers, onto coaches to co-ordinate diversions or to the departure gates to offer help and assistance to frustrated passengers. On days like these all the staff intergrated as one and unselfishly never considered their own arrangements - when there was a problem, work came first!

The staff working for Dan-Air achieved great job satisfaction, particularly when they could help others. I remember on Christmas Eve 1991, a young Australian couple arrived to take the last flight of the day to Paris, where they were to spend Christmas with relatives. Unfortunately they failed to obtain a visa, without which any Australian would be refused entry. Bryan Owen, the Duty Manager, swung into action. He called Immigration in France, but despite exhaustive efforts, was told that a visa was required and obviously it could not be obtained in time. Bryan approached the crestfallen couple and told them how wonderful Scotland was at that time of year. Before they had time to blink, he changed their tickets, booked hotel accommodation and they were on another Dan-Air flight to Aberdeen!

These sorts of incidents occured frequently and I received many complementary letters during my time as Airport Manager. Passengers often addressed letters simply to 'the girl with the long dark hair on the ticket desk'".

All customer contact staff received training from Customer Services Training, a department established in January 1992 under Senior Training Officer Brian Ablett. A programme of training was devised to build on existing strengths and experiences of staff so that even experienced PSRs gained much help from dedicated courses aimed at specific areas. 'Team Elite', for example, was the course for those working on the Class Elite check-in at Gatwick. These courses comprised of awareness of Dan-Air's scheduled services and charter products, as well as looking at 'good' and 'bad' customer service, the importance of 'first impressions' and ways of providing the right 'quality experience' for existing and potential customers. However, much was just common sense, good manners and having the right attitude, as Amanda Bubear explains:

"You had to be able to talk to the passengers, to chat to them and pass the time of day. We tried to demonstrate to them they were not just a number on a list. Often it was just little things like greeting the passengers by name instead of Sir or Madam. Whether at the check-in desk or at the gate it was easy for us to read their names from the ticket. It's amazing how many times you caught a look of pleasant surprise on their faces, for they were so used to the repetitive *'Good morning Sir... Thank you Sir...'* they get on some airlines.

Inevitably with the high level of security at all airports there were times when passengers did suffer from 'ruffled feathers'. Possibly the biggest, most regular cause of 'gripe' we heard time and time again occurred in connection with our international flights. So many passengers asked '...*you want to see my passport AGAIN?*' Passengers had to show their Passports at the check-in desk, at passport control before entry into the departure lounge and finally at the boarding gate. By the time they reached the gate they would have already put it away in their hand luggage twice and here we were asking for it again! We had to explain that it was required for security reasons, for in looking at the passport again at the gate we could then confirm that the name on the ticket was indeed the person that was travelling."

A NEW GENERATION - A NEW SERVICE!

There were many lighter moments to the job however, as Senior PSR Jan Hardy recalls:

> "Often there were incidents that raised a smile at the time and then were forgotten - they could help brighten a hectic time!. But I will never forget one incident... it was during the hurricane of October 1987. Things were getting quite hectic for we had many flight delays with numerous passengers arriving late at the airport due to the storm. I was on one of the check-in desks in the main concourse giving out tickets for meals because of the delayed flights. A lot of the passengers were anxious because they were late checking in and I vividly remember one lady approaching the desk and asking what was going on. I explained about the delays caused by the storm and she sincerely replied 'Oh dear, are there many trees on the runway?'. It was one of those occasions where you had to be there. I just could not speak, and all the check-in agents couldn't help any more people, they were falling about laughing!"

No matter how well trained, how adaptable and how properly motivated, the PSRs could not provide the right service to the customer without the right equipment for the job. With the increase in scheduled services in 1992 something had to be done to cope with the increase in passengers and their luggage, so other changes took place behind the scenes to improve customer care and at the same time make the airline more efficient.

Luggage checked in at Gatwick was speeded up by the application of new technology in a £35m upgrade of passenger facilities, part of which was the opening of 18 dedicated check-in desks during May.

Baggage tag production was automated and integrated with the computerised passenger check-in system in 1992, generating baggage tags in the same process as the passenger boarding card. This resulted in a new form of baggage tag that contained far more information than the old style. In addition to name, flight numbers and routing details, the tag carried codes that could be read two ways - infra red and a newly installed Optical Character Recognition (OCR) system.

As the passengers checked in their luggage, the new tags were attached to the bag in the normal way and then read by scanners which made use of technology similar to that used by supermarket check-out systems, feeding information to a computer system as the bags passed along the belts. The system then sorted the bags, and directed them to the correct aircraft.

Allison Beedie again:

> "The scanners could handle 60 bags per minute, which gave us ample capacity even at peak travel times. This system of computer OCRs both speeded up the transfer of bags and cut the chance of departure delays caused by bags arriving late at the aircraft".

Passenger requirements during a flight may have been looked after by the cabin staff, but the food, drinks and duty frees had to be organised to be on the right aircraft at the right time. Responsibility for this fell upon 'In-Flight'.

From the passengers' viewpoint, by far the most important was the food, and for almost every passenger carried on Dan-Air a meal was provided in one form or another. Responsibility for ensuring that the quality was to either the airline's or the charterer's standard was that of the Catering Department.

The staff within the In Flight section were constantly evaluating the meals to ensure that they met the passengers' requirements. The most popular meal was the hot breakfast served on early departures, but each menu type had four different cycles to ensure that regular passengers did not always get the same food.

New menus were introduced as a result of on-board surveys which revealed that passengers were becoming more aware of calories, so lighter and more health-concious meals were prepared. Any reasonable request would be processed by the catering staff, given enough notification. They handled everything from children's menus and ethnic meals to birthday cakes and flowers for special occasions!

Passengers had become more discerning over the years and as a result the standards in charter flight meals were re-evaluated. Cottage pie was out and vegetable lasagne was in! Nevertheless, there still remained a strong demand for the traditional British roast from holidaymakers returning to the UK.

The crew also had their own meal requirements from low cholesterol to vegan and Civil Aaviation Authority legistration dictated that two flight deck members must eat different meals.

Dan-Air uplifted food from 53 kitchens overseas which could become a logistical nightmare to control. Every menu had to be specified, photographed and priced and every piece of equipment from a teaspoon to a bar trolley had to be accounted for.

Duty -Free sales were also important to both the passengers and the airline and every year new samples of perfumes and products were evaluated, costed, negotiated and finally ordered to ensure that the customer was offered a wide variety of goods at the lowest prices. Part of this included the selection of wines for the meals, chosen to ensure a wide a range as possible to suit different palates.

Although all this back-room work could be hard on the staff, all took great pride in their work and enjoyed contributing to the end product - customer satisfaction - and seeing the passenger again on a re-booking!

DAN-AIR SERVICES

CHAPTER TEN
To the peak - and beyond

At the risk of some repetition, let us now look at the Company's most successful period. Broadly speaking this was from 1971, when the Group went public, to 1988.

The team was spearheaded in the seventies by Allan Snudden and Martin O'Regan, just 40 and 37 years old respectively, who had already amassed a tremendous amount of know-how in the business and Frank Horridge, the Deputy Managing Director, looked after the technical side. On the Operations side they were backed up by Bob Atkins and Arthur Larkman, whilst Bryn Williams ran the Engineering Division.

During this time the commercial side was growing fast under Frank Tapling, Errol Cossey and Don Siddaway, with a team that was second to none in the competitive world of charter business. The main activities centred around the inclusive tour market, where Dan-Air had excellent business relations with all the main tour operating companies in the UK and Berlin. Long-haul passenger charters to the USA and the Far East were flown in the summer, where Frank Tapling and Don Laws combined this business with freight contracts in the winter, which fully utilised the 707 fleet.

During the same period Dan-Air's Scheduled Services Division, comprising of the 748 fleet and some 1-11 400s, continued to expand under the direction of John Varrier, who was to join the Board in 1984. The large oil operations mounted from Aberdeen to Sumburgh and Scatsta mentioned earlier also came under his control.

In 1971 the turnover was £14 million, with profits before taxation of around £800,000, and the staff employed was 1,300. By 1979 the turnover had increased to over £125,000 with a profit before taxation of around £3 million, whereas staff levels had increased to about 3,000 people.

Right: Martin O'Regan, who was with the Company from 1967 to 1978.
Far right: Frank Tapling.
(DAS/GMS Collection)

DAN-AIR SERVICES

Right: Don Siddaway, who was with the company from 1963 to 1991.
Far right: John Varrier, who joined Dan-Air from Skyways in 1970.
(DAS/GMS Collection)

The break-up of a great team.
Unfortunately this highly successful team was broken up in 1978 when Harry Goodman decided to start up Air Europe, his own airline. Goodman took Martin O'Regan and Errol Cossey to the new company but, in order to limit the damage to Dan-Air, a compromise was arranged that limited further poaching to a few individuals and a three-year flying contract was agreed for a substantial amount of flying with Intasun. Dan-Air continued to enjoy a friendly working relationship with Intasun with substantial flying up until the summer of 1990.

One year later, in 1979, Allan Snudden left the airline to take up the position of Managing Director with the Luton-based Monarch Airlines and a new era commenced for Dan-Air. Frank Horridge, who had been Deputy Managing Director for some time, took over and Wilf Jones, Martin O'Regan's No. 2 became Finance Director.

It would be foolish to under-estimate the loss of three such capable personalities, but Dan-Air had the depth and strength within its staff to carry on.

However, looking to the future it was necessary to find suitable candidates for the top jobs and Graham Hutchinson joined in 1980, becoming the Deputy Managing Director until he succeeded Frank Horridge in September 1981, who then became Deputy Chairman. At the same time, Danny Bernstein was recruited from British Air Ferries and shortly afterwards took over the Commercial Department, where he remained until 1990.

In the Engineering Division, Bryn Williams, after 25 years with the Company, retired in 1980. His place was taken by Ted Evans, who joined the Board in January 1981 and for the next few years was supported by Ron Smith, who had been with the Company since 1958 and Len Crockford.

Gatwick Handling was continuing to be successful, establishing a new unit Manchester Handling in 1985 and later another at Stansted.

Ron Smith.
(DAS/GMS Collection)

TO THE PEAK - AND BEYOND

Mention should also be made here of Malcolm Davies, who joined the Company in 1972 as Personnel Manager. He also combined these duties with responsibility for Industrial Relations. Dan-Air always enjoyed good relations with the numerous Trade Unions and Malcolm Davies had many a long negotiating session with their representatives and always managed to strike a balance that was acceptable as a compromise to both parties.

During the early 1980's there were further changes to the Board, with Michael Newman and David Herbert being appointed. In 1986 Sir Ian Pedder became Deputy Chairman, prior to becoming Chairman in 1988.

This was the period of maximum growth. Turnover in 1980 was £150 million with profits of just over £2 million and a staff of 2,900. By 1988 turnover had risen to £335 million, with profits of nearly £10 million. During this time asset growth had also been achieved by prudent management.

With Dan-Air Services responsible for 95% of the Group's turnover it was natural that the Davies & Newman Holdings Board should exercise tight control on its main subsidiary. However, the Airline Board was responsible for its own day-to-day affairs, but monthly meetings of the Holdings Board supervised the Airline results and authorised large capital spending.

Newman House was the adminstrative centre at Horley. This custom-built complex was opened in 1980 by Norman Tebbit MP.
(DAS/GMS Collection)

DAN-AIR SERVICES

Events in the industry.

In the eighties Dan-Air was also playing an important role in the 'politics' of the industry. This was not only on their own behalf, but also on behalf of the other independent carriers who were striving to maintain their position against nationalised companies, both at home and abroad.

The Department of Trade, the Civil Aviation Authority and the European Community were all formulating new policies on such matters as licensing, flight time regulations, pilot duty hours, airport slots and many other matters. At home, the privatisation of the state airline British Airways was a big issue, which was likely to have a profound effect on the independent carriers.

It is interesting to see the Company view-point on the matter. As early as 1984 Mr Newman was informing shareholders at the AGM with prophetic words as to what was going on behind the scenes:

> " As is well known, the Government's intention to privatise British Airways had led to much speculation about the effects this will have on independent airlines. The Civil Aviation Authority, at the instigation of the Transport Minister, has asked airlines and others to present their views on this and other policy matters affecting the industry. About 100 submissions have been made and a wide variety

The 1988 Board of Directors of Davies & Newman PLC outside the London Headquarters at New City Court. Seated (left to right) Sir Ian Pedder KCB, OBE, DFC, Mr J. W. Davies OBE, Mr F. E. J. Newman CBE, MC, Mr D. P. Herbert, Mr E. G. Hutchinson. Back Row (left to right) Mr. D. J. Quinn ACIS, Mr E. C. Hartwell, Mr M. R. F. Newman, Mr. W. Jones, Mr. D. L. Bernstein, Mr R. Payton. (DAS/GMS Collection)

TO THE PEAK - AND BEYOND

of views have been expressed...
It is impossible to know at this stage to know what changes, if any, are likely to take place but provided they are reasonable Dan-Air is ready to accept the challenge in whatever way is necessary".

The following year Mr Newman brought the shareholders up to date in his statement to the AGM:

"Overshadowing the year's operations was a wide-ranging consultation with the Civil Aviation Authority and the decision taken by the Government with regard to the future of the industry in the light of the privatisation of British Airways. Considerable disappointment and perhaps some disillusionment at the outcome was experienced as a result of the Government endorsement of British Airways' dominant position which in some respects has been strengthened.

The rejection of the Civil Aviation Authority's review of the Industry was a severe blow to the private sector as well as harmful to the Industry and in the long term may lead to less competition than more. This decision, together with a lack of decision on a number of other matters, notably the provision of adequate airport facilities in the South-East, has led to further uncertainty in the industry. It is to be hoped that on reflection, the Government will see the wisdom of leaving the licensing policy to the Civil Aviation Authority who are a long-standing independent body, rather than become embroiled in the day-to-day arguments revolving round these issues".

Industry-wide events were still being mentioned by Mr Newman at the 1987 AGM:

"Following the privatisation of British Airways, the British Airports Authority will follow them into the private sector if the present Government has the opportunity. The provision of adequate airport facilities in the areas where they are most needed is the number one priority to ensure continued growth in traffic. This should be provided ahead of the requirement rather than as a last resort when over-crowding of runways, terminal buildings and airspace become intolerable. Unless Governments take the matter seriously, in particular with Local Authorities, the advantage that Great Britain has as a staging post for Europe on behalf of the long-haul travellers will be lost".

*Mr J. E. J. Newman MC, CBE, the Chairman and Managing Director of Davies & Newman PLC poses in front of BAe 146 'HH..
(DAS/GMS Collection)*

Much time was spent during this period on these vitally important matters and the Board was greatly assisted by Mr Bailey, who had recently retired from the position as Airport Director at Gatwick and Stansted. In the Company, Peter Somers was Head of Department responsible for External Affairs and was also heavily involved in industry matters. He was Chairman of the Gatwick Scheduling Committee from 1970 to 1993 and Chairman of several Committees relating to air traffic flow management. Both Pat Bailey and Peter Somers played a vital part, both in the Company and in the industry as a whole.

1988 had been the year of peak achievement, but towards the end of 1989 the Board had become aware of the changing nature of the airline business. Pressures were being felt by all carriers and the outlook for the future for small and medium airlines was looking bleak.

The Board, therefore, decided to take steps to change direction; either by seeking a partnership or by placing Dan-Air in the hands of a company with greater resources.

A number of negotiations were carried out over the next twelve months, but deteriorating results and complicated regulations restricting overall control of airlines by foreign companies negated all efforts to find a solution.

DAN-AIR SERVICES

CHAPTER ELEVEN
So near disaster...

A world-wide down-turn in the airline business affected every airline operator during the late 1980s Nowhere was this more strongly experienced than in the charter companies that were concentrating on the highly competitive European markets. Dan-Air had carried over six million passengers in 1989, but despite this, the airline made a loss of £3.34 million.

By the end of the year, the Company announced that it was disposing of the two high capacity 336 seat Airbus A300s in order to recoup a half-year loss of £7.6 million. At the same time a third 737-400 would join the fleet in the spring of 1990 as part of a move towards fleet standardisation.

During 1989 Dan-Air was still very busy, but increasingly high interest rates were causing tremendous problems for many of its customers, who cut back on their travelling plans; thus the Company was beginning to suffer from a severe cash shortage. Historically, business (and therefore cash-flow) had always slowed down at the end of each summer season. In these annual lean winter month, an amount of prudent belt-tightening always occurred until the money from next summer season started to flow into the coffers.

The worsening market conditions for aviation during the second half of 1990 reflected international concern at the onset, in August, of the crisis within the Gulf, coupled with severe recessionary factors. 1991 proved to be a very difficult trading year: the problems in the Middle East escalated into the Gulf War, which, although over very quickly compared to the predictions of some pundits, had a severe and prolonged effect on the plans of the world's travelling public. Nowhere was the effect felt as severely as in civil aviation.

But there were also other forces at work. Throughout September and October of 1990 Harry Goodman's International Leisure Group attempted to deal Dan-Air a fatal body-blow by disparaging the airline in the press on one hand, and simultaneously withdrawing the engineering contract for the Air Europe fleet from Dan-Air Engineering and all of Air Europe's charter work away from Dan-Air. This had the instant effect of turning Dan-Air's cash-flow troubles from a problem into a crisis of major proportions.

In order to be able to understand more fully what was going on, and the implications of each 'action' and 'reaction', we must now leave the Jet1-A perfumed atmosphere of airliners, ramps, routes, services and people that made up Dan-Air and enter into a totally

DAN-AIR SERVICES

different sphere - the rarefied world of statistics, market analysis, share issues and high finance.

The increasing tension in the Middle East and the deepening recession had an immediate effect on many businesses which then severely restricted staff travel arrangements, effectively removing one of the most profitable revenue contributors to the airline. Similarly, the winter leisure season began just as the Gulf crisis reached its climax and it became clear that the crisis was about to turn into a shooting war. This brought a further substantial reduction in both business and leisure passenger traffic.

Potential disaster looms...

It had already become apparent by the middle of 1990 that the level of charter work available to Dan-Air had reduced so rapidly that it could not be compensated by the growth of the scheduled service network. As a result, the Board identified that Dan-Air would have insufficient working capital resources available to enable it to continue to trade through the winter of 1990/1 without a large increase in its level of bank borrowing facilities.

In the summer of 1990 the Board of Dan-Air had a number of potential options open to it to rescue the Company. Throughout June and July the Board was very close to selling the airline to British Midland; at one stage terms were agreed and the parties were within a few hours of doing the deal, but it was aborted. The Board was also close to selling to British Airways during that time through the offices of Baring Brothers & Co Ltd (Davies and Newman's financial advisers) but that also came to nothing. They had also established a dialogue with David. N. James, the renowned 'Company Doctor' who had built up a repuation of saving companies that found themselves in difficulties.

James' early career was in banking and finance, including a ten year term (concluding in 1973) as a member of the founding management team of Ford Motor Credit in the UK. Subsequently he joined The Rank Organisation PLC and, during an eight year period up to 1981, was successively managing director of three financial and industrial subsidiaries. In recent years Mr James had undertaken a series of corporate rescues, including Central & Sherwood PLC, North Sea Assets PLC and Eagle Trust PLC. David James takes up the story:

'Company Doctor' David N. James CBE. (DAS/GMS Collection)

"The Board had a dialogue going with me (instigated by Barings) for the possibility of a rescue since 12 July 1990, but I was regarded as the 'back-stop'; they would sensibly have preferred to have sold the business as a going concern then, but when they failed to do so, they faced the need for re-financing. The Board decided that their case would be helped by new executive management; Barings then recommended the Board to invite me to go in to lead the re-financing.

SO NEAR DISASTER...

*Tails to the fore!
A pair of A300s,
G-BMNC and 'MB with
Company 727 "November
Foxtrot behind at
Gatwick during April
1988.
(DAS/GMS Collection)*

I set up the structure of dialogues for the Board to negotiate the re-financing and had gone off on another affair to Los Angeles. I was there on Tuesday 16 October 1990 when, very early in the morning, I received a telephone call from Fred Newman and his advisers. They said they were in the middle of a meeting with the banks and had been offered terms. The bank had stated that this was the only basis in which they would proceed and if the Board accepted those terms, I was now being asked, would I equally agree to be Chairman?

I refused, for I thought that the terms the bank was asking (a special fee of some £15 million) were outrageous. The Board told me that they could not shift the bank and I told them that I was refusing to take the job. They replied this meant that the Company would have to fold - it could not go on.

I offered to fly back that evening and arrived in London during the early afternoon of the 17th and went straight to a meeting with the directors and Barings. As a result I got the banks to agree to a further meeting at 2 o'clock on the Thursday afternoon, the 18th. This was going to be a last ditch attempt, for we had to issue bonds by the close of business (3 o'clock) on Friday 19th. If we had not been able to issue those bonds, the airline would have had to fold that day.

It was decided that I would go as representative of Mr Newman, I would take my own advisory team and that Fred Newman and other Directors would be available at New City Court either to comment on or endorse what I could negotiate.

I went in and negotiated the special fee down from the original £15 million to the new minimum £8.75 million and improved the terms of the transaction substantially. That meeting took eleven hours and five minutes of non-stop negotiations. We finally shook hands at 1.05 a.m the next morning and announced the deal at a mid-day press conference, so we effectively saved the airline with three hours to spare.

DAN-AIR SERVICES

In announcing the arrangements for re-organising the company at that press conference on 19 October 1990, David James said *"There will be no fire sales and no bargain deals. All next year's charter holidays which have been booked are absolutely sound"*. The press announcement stated that he had persuaded Lloyds and a number of overseas banks to provide the £70 million needed to cover the Company until at least December 1991 and promised that most of the airline's 3,600 employees would keep their jobs.

Immediately on completion of negotiations, a major re-organisation of the Board took place. On 31 October 1990 David James succeeded Mr F.E.F Newman as Chairman of the Board. The Board, under Mr Newman, had already accepted that redundancy conditions would apply to J. W. Davies, E.C.Hartwell, W. Jones, and D. L. Bernstein arising out of the transfer of the Company's Head Office from New City Court to Newman House.

The situation was by no means rescued by the winning of the new banking facilities, as David James explained during an extended interview with the author in January 1993:

"I came in with a new facility of around £40 million, but I very carefully obtained from the banks an understanding that I would be allowed to re-submit an analysis after twelve weeks of my time in office. When I had done my own analysis I was then able to address the level of working capital really needed. In this case I found that we were £12 million short of what was required but I was able to get the extra facility from the banks."

The engineering side had been greatly affected by the overall downturn in business as a result of the looming threat of an almost certain war in the Gulf area. The Board had invited tenders for purchase of Dan-Air Engineering (DAE) previously, but they had been suspended. David James put out a new tender document through the offices of Barings and a number of offers came in. The Board thought the positive reasons for this disposal were many-fold for it was having a dire effect on the core business - the airline: DAE had a capacity some 45% beyond that of the airline's own in-house requirements, there was under-utilisation of DAE due to financial constraints and the past investment in the Division's

The recently completed wide-body sized hangar complex at Gatwick that was sold to FLS Aerospace, with Airbus November Charlie parked in front. (DAS/GMS Collection)

SO NEAR DISASTER...

assets would be released. The facilities that the Division had were good; one unique feature was the only 747-sized hangar at Gatwick not owned by either British Airways or the British Airports Authority, the Division also had a high reputation within the industry. Thus the Board had 'live' interest coming from a number of potential purchasers. The best of them, from the Danish-owned FLS (Aerospace) Ltd for £27.5 million was accepted, but nevertheless, the Board faced a terrific battle to process the paperwork to complete the sale before conflict in the Gulf broke out. Christmas and New Year holidays were cancelled, and finally the contract was signed at just after 3.20 a.m. on 15 January, only hours before the shooting started.

The final transfer of ownership to FLS took place on 28 February 1991, thus reducing the overall staff levels of Dan-Air from around 4,300 to 2,500 and injecting a massive inflow of funds to enable an increase in banking facilities.

Even so, for the six weeks of the war, Dan-Air suffered an outflow of funds that amounted to £28-29 million. David James brings the picture into stark relief:

The Domestic Timetable for the winter of 1990/1 (28 Oct - 30 March)

" On Friday 8 March (the day that Air Europe folded) I had £920,000 left of the facilities I could draw down from the banks. That was all I had left, before we would have gone the same way, despite the fact that we had sold DAE for £27.5 million a few weeks before. From that moment on our cash-flow picked up, but we were almost as close as Air Europe was to collapse."

He also thought that the problems with Air Europe had not been fully perceived early enough:

"Although now I enjoy a very good relationship with the CAA, there were times when they were very 'sticky' towards us. I formed the view very early on in my time with Dan-Air that Air Europe was experiencing very severe problems. I was somewhat fazed therefore, by the realisation that the regulatory bodies did not seem to perceive the hazard with them, but were very aware as to what our problems were. As a result of this I felt we were getting a slightly raw deal, in that we were being put under a great deal of pressure that I did not see being applied to Air Europe. I was not in the least bit surprised when their problems surfaced and it became clear that they had a far greater terminal risk of collapse than we did.

I felt that the CAA were very hard to deal with during the first half of 1991 and they did not really expect us to succeed with our re-financing in the autumn of 1991. During that time we had been under some considerable threat as to whether our licences would be withdrawn."

Boeing 737 - 400 November Lima unloads its passengers at Nice. (DAS/GMS Collection)

183

DAN-AIR SERVICES

There had always been some degree of staff communication within the Company, but because of the momentous changes happening within the Group, a policy of greater employee communication was established. This took the form of a series of 'Chairman's Newsletters' distributed to every member of staff as items of news became available on the progress of the Company. For example, in 'Chairmans Newsletter 5' dated 5th March 1991, David James starkly reported the options:

> "In my comments to Shareholders at the EGM four weeks ago I said we probably had three seperate strategic alternatives to consider. Basically these were: seek to refinance the airline but carry on independently in our present structure; merge with another airline to become a larger, securely financed unit; sell Dan-Air to another airline. Realistically, it is the first two of these alternatives which justify the most urgent consideration although it is still too early to offer a forecast of the outcome. All I can say is 'hang on in there'. I will advise you of the progress as soon as possible".

Six weeks later, In 'Newsletter 6' the die had been cast:

> Our objective is now to establish Dan-Air as an independent, stand alone airline. I believe that we can substantially achieve this with the available resources..."

As James has already explained, the conclusion of the Gulf War brought about an immediate recovery in both business and leisure travel, but the improvement in March 1991 did not maintain its pace into April, although a trend towards more normal levels of activity did slowly became established.

This photograph of BAe 146-300 G-BPNT shows the rear fuselage extension for more passengers to good advantage. The airbrake built as the tailcone is also clearly visible. (DAS/GMS Collection)

184

SO NEAR DISASTER...

A further disposal, this time the ship-broking activity carried out by Davies and Newman Limited, occurred on 25 October 1991 when it passed to a subsidiary of the London Airtaxi Centre Ltd, a concern owned by members of the the Newman family. A further disposal occurred on 8 January 1992, when Dan-Air completed the sale of its shares in Manchester Handling Ltd to Gatwick Handling Ltd and Aer Lingus UK Holding Ltd. As a result, Manchester Handling became owned equally by the two companies.

Securing the future...

Before setting about developing a detailed financing plan to attempt to secure the Group's future, a closely focused strategy had to be established. The broad features of this were to:

a. Concentrate on Dan-Air's operation and development as an independent airline.

b. Develop the airline as a specialist short-haul scheduled services provider within Europe and the UK based upon the Gatwick Hub.

c. Maintain subsidiary scheduled services spoke centre based at Newcastle, Manchester and Berlin.

d. Progressively modernise the fleet of aircraft.

e. Continue a strong presence within the charter market, albeit with a reduced but more flexible capacity than previously.

In formulating this strategy, the Board had to consider the feasibility of a return to profit based on a substantial transfer of marketing and operating emphasis from Charter to Scheduled Services. In order to arrive at this decision, the Board took into account the following points:

i Dan-Air had an established and substantial presence, and was at that time was the only Gatwick-based scheduled service operator that was committed to short-haul operations. Although competition might develop, it represented a strong market position to build on.

Boeing 727 G-NROA (DAS/GMS Collection)

DAN-AIR SERVICES

146-300 G-BPNT just after getting airborne. The fuselage extension of this variant when compared to other versions is particularly noticable (DAS/GMS Collection)

ii Dan-Air's cost structure reflected a highly efficient operation that should help the airline to achieve a better profit contribution on short-haul routes compared to larger competitors.

iii Upgrading the fleet would bring substantial operating economies and enhanced flexibility thus improving general levels of efficiency and profit margins.

iv More flexible charter capacity would help level out the summer peak and winter trough effect that had traditionally created problems in financial planning.

v New scheduled services could be added to achieve progressively higher direct contributions to profitability.

Thus by June 1991 the Board had arrived at the conclusion that it was feasible for the Company to return to profitability provided that new sources of capital became available to support this planned development. Further steps to re-structure the Company occurred in September, when the announcement was made of the appointment of Vic Sheppard to the new position of Director, Customer Services. Vic Sheppard (who was later to leave to establish his own airline called, at the time of writing 'First European') had been with Dan-Air since 1988, after a period with British Caledonian. It was Vic Sheppard who had been responsible for the highly successful launch of 'Class Elite'.

1-11 G-ATPL loads up with passengers at Tarbes Ossun Lourdes (DAS/GMS Collection)

SO NEAR DISASTER...

Just after this, the sales and marketing of both scheduled and charter services were brought together under the control of Charles Powell, who summed up the objectives at the time as follows.

"The streamlining of management will provide a more aggressive marketing thrust and better quality service to travel agents and tour operators. The re-organisation will maximise the undoubted abilities of staff and speed up decision response time from agents. The 'one airline' concept will now mean that where agents and tour operators previously had to deal with two departments, they will now deal with one. The reorganistion enables a personalised approach to agents and provides a better selling capability allowing the industry to provide a more efficient service to the general public. The internal re-organisation will also strengthen the front line-sales team who have day-to-day contact with the travel industry".

A change of emphasis

In the autumn of 1991 a successful equity issue raised £53.75 million through the issue of 107,500,000 new ordinary shares, with 750 members of staff demonstrating their commitment and identification with the Company by becoming shareholders under the terms of the offering. For the purpose of putting this together, David James had gathered a team of top City advisors to work on this share issue. The team comprised: Coopers & Lybrand (who took a leading role in the essential due diligence process, including a review of the illustrative projection as to future profitability), County Natwest (who had provided considerable support and who had brokered the deal with the City); and Barings, who were the signatories to the circular endorsing the recommendation to the shareholders.

The re-styled check-in desk line at Gatwick, showing both 'Class Elite' and standard ticket facilities.
(DAS/GMS Collection)

DAN-AIR SERVICES

In order to achieve the 're-birth', David James won the support of nineteen top share-holding institutions, led by Schroder Investment Management, for the re-financing with a four-year illustration of profits that demonstrated a switch from losses of £35 million in 1991 to profits of £42 million in 1995. Of all the institutions approached, only one rejected the idea, and this was due to internal conflicts of interest.

The profits turn-around was to come from a shift away from charter flights towards scheduled services. Dan-Air also announced plans to replace some of the older aircraft it operated by leasing 20 new aircraft over four years, thus reducing the average age of the fleet from over 14 years to a little over 5 years.

Integral to this was the winning of at least seven of the eleven European scheduled service licences it had applied for, some of which had been operated by Harry Goodman's recently collapsed Air Europe. However, the additional CAA licences needed to operate these routes could not be granted until the Company was able to demonstrate an adequate financial stability.

The Board felt that they could now reveal the plans for future development in greater detail. The fleet operated would be progressively modernised, with a target set to reduce the average age of the aircraft to 5.4 years by 1995. At the same time, it would standardise on just two aircraft types - the Boeing 737-300/400 and BAe 146. A greater emphasis would be placed on scheduled services total activity, increasing its contribution to turnover from 34% to 62% by 1995. At the same time the airline would remain a major charter carrier, but would improve the quality and profitability. The size of the charter fleet would be reduced to 15 aircraft to address the problem of poor winter utilization, over-capacity and low yield.

A major advertising campaign was launched to back up the latest moves, commencing on 27 January 1992. The main thrust was concentrated on projecting the message that Dan-Air was the largest airline operating out of Gatwick, consequently the theme was to draw the easily identifiable connection between Dan-Air

Boeing 737 - 300 G-SCUH in the turn. (DAS/GMS Collection)

SO NEAR DISASTER...

and the expanding network of European operations.

Unfortunately, this re-financing and advertising thrust was followed by a further period of market uncertainty in early 1992 which preceded the British election. This had a depressing effect upon passenger traffic from mid-February into April.

By the middle of 1992 Dan-Air had achieved much of its re-structuring strategy following the near-collapse of autumn 1990. The Group had concentrated on developing Dan-Air Services Limited, with the objective of progressively extending its capability and performance as a scheduled airline.

In the expansion of its scheduled service routes, the Company was successful with all seven of its applications to the British Civil Aviation Authority (CAA) for new licences on its preferred routes, commencing operations to five new destinations - Oslo, Stockholm, Athens, Rome and Barcelona. The CAA granted licences to Cairo and Istanbul also, but the airline was not successful in gaining the requisite approval from the countries concerned. Nevertheless, negotiations continued, with the intention that operations would commence on these routes as soon as was practicable.

By May 1992 Dan-Air was operating a scheduled network that served a total of nine U.K. destinations and twenty-one in Europe, circumstances that allowed it to consolidate its position as the United Kingdom's second largest airline with a scheduled network when measured by revenue-passenger-kilometres.

The Company had access to statistics from the British Airports Authority, from which they were able to measure the entire scheduled market for the south-east of England, covering Heathrow, Stansted and Gatwick. From these figures, it could calculate the actual growth in all passenger traffic achieved by Dan-Air for the first three months of 1992 compared to the previous two years.

The figures showed that overall in 1992, the scheduled market was 19.4% above 1991. Dan-Air had a comparable growth over 1991 of 21.7%. For 1992 growth over 1990, the industry figure was only 1% whereas Dan-Air advanced by 25.7%. Obviously this massive increase reflected the Company's major commitment to the scheduled service sphere.

There was strong evidence that Dan-Air had also increased its share in key markets over the same period. These advances were particularly strong on routes to France, with Montpellier up from 65.8% to 96.6% and Toulouse from 36.2% to 47.0% The Gatwick - Paris route, despite strong competition, was holding a rising share of 11.9% of London/Charles de Gaulle traffic.

There were other high-performing routes. Zurich was up from 14.8% to 17.9% and Jersey from 32.6% to 39.5%. The success was

Nichola Ashworth models the near definitive Dan-Air uniform. Numerous different blouses and scarves were worn at times to create a different appearence, but the only major change after this was the hat, which was not liked by many, was dropped.
(DAS/GMS Collection)

189

DAN-AIR SERVICES

not confined to overseas routes; the Gatwick - Aberdeen service was up from 15.7% to 18.5% and Gatwick - Manchester was holding steady at 18.5%

However, there were cases where the market share had fallen, usually due to special and predicted factors. The Gatwick - Berlin route had fallen from 15.5% to 7.9% as Tegel Airport was opened up to German airlines, thus Dan-Air reduced flight frequencies. The Gatwick - Nice service also suffered a fall, from 14.1% to 12.3% due to the introduction of two new carriers as competition.

Overall, the analysis supported the Board's continuing confidence that the Company would achieve its objectives, despite the difficult market. However, as David James said in his address to shareholders at the AGM on 29 May 1992

"...Given the volatile and fluctuating pattern of the market so far in 1992, it may be appreciated how difficult it is to arrive at a confident assessment as to the likely out-run from the 1992 year overall. Even so, it must be said that the forward booking pattern, to which I referred earlier, for the peak revenue earning months of the summer, does indicate that a return to overall profit in the 1992 year would now require a further and sustained upturn in both passenger traffic and yields above that currently in evidence."

By comparison, the charter side of operations had been substantially reduced to a total of fifteen aircraft, all of which were of enhanced quality. The airline felt that with this target achieved, a stabilisation on this figure would allow for a much improved balance with the fluctuations of seasonal demand.

Many of these events came into force on 18 May 1992, when the airline undertook the largest single route expansion in the Company's history. The Gatwick - Barcelona service came on stream, the Vienna and Oslo services were doubled to twelve per week, the Berne route changed from weekend only to daily, with 'Class Elite' offered (the first and only business service on this route) and flights to Brussels increased to four times a day.

Further to this, a major strengthening of the airline's Sales and Marketing Division occurred with the creation of three new senior positions. In the summer of 1991 Charles Powell was appointed as Sales and Marketing Director, in June 1992 John Fitz-Gerald was appointed UK Sales Director from Cathay Pacific, Robin Wicks joined the team as Head of International Sales and Ian MacDougall took the position as General Manager Revenue Performance.

At the same time a new internal appointment was made in order to exploit the rapidly growing importance of Gatwick as an 'interline' centre (a place where passengers change from one flight to another, which could be Intercontinental to International or Domestic or vice versa).

During 1992 a campaign was started in the press to make the company more of a 'personal choice' airline...

SO NEAR DISASTER...

Brian Brett was appointed as General Manager, Airline Relations and Overseas Sales.

John Olsen, Dan-Air's new Managing Director and Chief Executive, said at the time...

"New routes, the stepping up of services and the arrival of the first six new Boeing 737's of the fleet modernisation programme are very positive indications of Dan-Air's planned progress as a major scheduled European airline. The addition of these key positions adds great experience and skill to our sales and marketing team. This added strength will enable Dan-Air to develop and exploit still further our market position in all areas of business".

John Olsen had joined the Swire Group of Companies in 1966 and had undertaken a number of assignments for them based in London, Japan, Southeast Asia and Hong Kong. He had been appointed to the Board of Cathay Pacific in 1985 as Commercial Director and had also been the Chairman of Hong Kong Air Terminal Services, Securair, Securair International and had chaired the Executive Committee of Swire Air Caterers Limited. He was General Manager, Europe of Cathay Pacific before joining the Company as Chief Executive in April 1992.

The author's extended interview with David James in January 1993 sheds more light on what was going on in the marketing side and the reasons from bringing in John Olsen. David James:

"John Olsen brought with him a team of absolutely died-in-the-wool, hard nosed marketing people with lots of airline experience that 'shook up' what I saw as the Dan-Air malaise. The Company had never really succeeded in getting into the groove of marketing itself with the philosophy of a scheduled airline.

We had marketed ourselves as though all the flights were charter. There was a classic example of the problems that this could cause...

Charles Whyte did a marvellous job from a zero start point in beginning to get into place a system whereby we could analyse the forward loads on each flight and the mix of each different category of fare. We never got to the point where British Airways were (they could monitor about a year ahead) but we did get to the point where we could get what we thought was an accurate reading on each flight and each route so that we could see how many seats we had to sell for at least three months beforehand.

Unfortunately we had some serious misreadings in this. For example, on the Zürich route, I would look at the figures and see that we were due in a given month to fly 15,000 passengers. I would see that with eight weeks to go we had already booked 12,000. On seeing that I would be saying 'Marvellous, we're going to hit the budget and go way beyond for that month'. But when the month came, we had probably only got another 500 bookings, well short on what we wanted for the month. I could not work out what was happening, but then I caught on what was going on. We were selling seats in blocks to discounters and consolidators instead of building the momentum of being an individual choice airline to be picked by the customer. We were selling too much on the block booking basis as though our scheduled services were really 'charters in little packages'.

John Olsen's team made a tremendous impact once they started work. The whole graph of our performance went way up. The number of seats sold in the four weeks up to the month rose dramatically and the longer the team worked, the more late bookings came in. These brought in better yields, for we were now selling to individuals, not organisations that demanded a discount. From June the scheduled services shot up - we monitored daily the new booking figure and the actual flown figures for the previous day and they were rocketing ever upwards!

... this marketing campaign, along with other tactics brought an ever-increasing level of bookings...

DAN-AIR SERVICES

The same could not be said about the performance of the charter side. Something went drastically wrong in the beginning of June 1992, and there are still differing views as to what it was. My favourite (and I will only know for sure when tour companies publish their results) is that they had geared up for far more capacity for 1992 than the market actually justified. They caught a cold when they saw the first-half year figures and so decided to cut capacity. Dan-Air were the marginal swing provider, so they cut us out. The tour companies gave the entire priority to their own in-house carriers with the view that the independent providers such as ourselves could go to hell. The second aspect to that is that I think that they have had a far worse year than they have yet admitted to and they were panicking to cut their own costs at that time. Thirdly, I think that many companies with dedicated carriers were under very great pressure to load as many passengers as possible in their own fleet in order to appear to sustain the trading value of their aircraft.

In terms of nett revenue turned into cash I would estimate that about 10% of the charter work got cancelled in June. This was followed by a 20% reduction in July, 25% in August... The net revenue effect was about £14 million out of our cash.

By the end of July the sales figures were not presenting a very rosy picture. In 'Chairman's Newsletter No. 12' David James summed up his feelings: *"Bluntly, the 1992 market is a dog"*.

As a result, the Board recognised that the Group would not be able to generate the reserves of cash required to keep itself going through the winter of 1992/3. Therefore the Board urgently sought some kind of alliance with a strong airline partner. Talks took place with major airlines in the UK, EC, USA and Far East but failed to achieve a satisfactory outcome.

By August discussions between the bankers and the Board confirmed the need to identify a new source of capital if the Group was to trade beyond the third week of October. Accordingly, the Board prepared a further rescue package proposal to present to its shareholders. Talks took place with a number of principal shareholders who expressed some support, but thought that this would fail to achieve the required level unless some industry alliance could be concluded at the same time.

...for the scheduled services, selling both standard and 'Class Elite' seats...

Travel in space. Fly Class Elite.

DAN-AIR

Speculation of a Merger

During mid-September 1992 there appeared speculation within the media, with rumours flying around of a possible merger between Dan-Air and Richard Branson's Virgin Atlantic airline. There had also been speculation over a possible 'marriage' with either American or United Airlines from the USA, so as to create a ready-made 'European' arm to their operations. Over the weekend of 26/27 September this intensified with many rumours which culminated in suspension of Davies & Newman shares on the Stock Exchange at 0740hrs on Monday 28th, following the announcement that negotiations were taking place between Richard Branson and Davies & Newman. David James explained the reasons:

SO NEAR DISASTER...

"We decided on Friday evening that we would request the London Stock Exchange to suspend the shares of Davies & Newman. The reason for this is that we have a responsibility to ensure that no false market is created in our shares based on the expectation of the value of the agreement..."

The deeper facts that lay behind the public statements reveal just how involved and complex the picture had become behind the scenes. David James again:

"We talked altogether with about 20 industry players. It is normally my custom when I go into a new business to make contact with the major industry personalities anyway, and so I got a dialogue going with many right from the beginning. There was a wide level of interest in the dialogue with Dan-Air, but there were groups of specific interest. For example, initial dialogue with most of the American airlines always came up with the same concern - we were in the wrong airport. Even when that was not the case, there was still the major concern than none of them would touch us until they had succeeded in kicking the British Airways/US Air deal into touch. They did not want to create a reciprocal deal which could be turned by BA against them as evidence that BA should be allowed to gain a foothold in the USA in return for the Americans taking a similar presence with us. So, in many respects, the BA dialogue over US Air killed any prospect we had of getting a deal going with the American carriers.

Secondly, when the European sphere was viewed, you had Air France who were not enamoured with Gatwick, but which lost out with a number of initiatives, including TAT, which has gone to the BA camp. At one stage, we suggested that they should come in with us, under the code-name 'Tit for TAT', but we could never conclude a deal.

We had started on 1 July 1992 to produce the 1993 business plan, which was presented to me on 4 August. We knew that we would have to address a re-financing of the Company for the 1992-3 year, even if it was only a question of bank facilities. That plan showed that, in our view, it was viable for the Company approximately to break even in 1993, even on the assumption that there would be nil growth in the marketplace, nil GDP growth in the UK economy, that the yield would only improve at the level of price index inflation and that there would be no real growth in passenger numbers. We could see that, provided we could find a way of getting the Boeing 727s out of the books as soon as possible (which had been a tremendous drain as they were older, less fuel-efficient aircraft) there was even a chance that we could actually make a profit.

...and did much to put Dan-Air, and Gatwick Airport in the public's mind...

We had negotiated at that time - and it had been agreed - for the sale of the entire 1-11 fleet to the International Leasing and Finance Corporation for what would have amounted to a book loss of about £6 million. We were then going immediately to short-term lease back six of them until the end of the 1993 season and then accelerate the replacement of the 737 standardisation plans. The cost benefits of that alone were tremendous! Also we would have got standardisation of equipment and interchangability on all the routes to a far greater level. This was in addition to the deal we had done with British Aerospace where we had arranged for them to take our old two 146-100s in return for putting in -300s and give us £5.2 million book profit in the process.

All this amounted to a major reduction in fleet size, but we thought that the scheduled services could work on a far higher aircraft utilisation rate (as a result of John Olsen and his team's work), by standardising on the 737. It would have meant a smaller workforce, but as many as possible would have been re-trained and integrated.

After we had completed our plan, we asked Coopers & Lybrand, County NatWest and Barings to start due diligence with a view to assessing that plan as a means to start raising new capital. I had met with all our major shareholders during July and told them what the state of things was, and told them that we hoped to get through without any further capital, but that we would not know until the autumn.

No sooner had we started validation of our plans than the value of sterling plummeted. We had some £36 million of outgoings which were exposed to

DAN-AIR SERVICES

the dollar; the immediate impact of the collapse of sterling took £11 million off the bottom line of the Company. Before we had been looking at between break-even and a £5 million loss; we were now looking at a loss of between £15 - 16 million for 1993. I knew now we had no alternative but to look for new capital.

We finally reached a point where our advisers were prepared to take us to the institutions again to see whether we could raise the money - a total of some £35 million - and we were sure that if we got that, the banks would give us the extra cash. Those two factors together would give us adequate headroom to clear the £55 million we would have to cover in May 1993. It was also possible new cash would allow me to fulfil my ambition to raise the extra £10 million to attack the 727 problem. On that basis I felt that we would have a survivable airline that would take us through until at least the end of the 1993/94 winter season. Therefore we would have a two year absolute built-in security of survival in which to get the marketing and balance of the airline right.

There was much talk as to the problems Dan-Air was experiencing in relation to cyclical cash-flow. Many people failed to understand fully just how stark that problem was. For example, had the Company retained its independence the following figures would have applied. In September of 1992 we had something in the region of £5 million on deposit. In contrast, our peak of borrowing in April/May 1993, would have required at least £55 million to get us through.

We went out on 24 September and saw five institutions that aggregated 52% of the total equity holding of the Company. At the end of the day, those institutions indicated that they would be prepared to put up what amounted to about £12 million. They were reluctant to put up any more unless we could assure them that we would form an alliance with another airline.

Well, if you've only got £12 million out of 52% of your shareholders you're not on at a strike rate that will give you a total of £35 million from 100%. At a Board meeting in the offices of Barings that started at 6.30 that evening we had to decide if we were to go on and visit another half dozen institutions the next day or recognise the fact that we had failed and consider inviting the banks to appoint a receiver.

At exactly 7 o'clock there was a call from Richard Branson saying, that he would very much like to do something to rescue the airline; he wished to become an equity player and could we talk immediately? Richard's great interest in us was really as a means of interlining and feeder distribution for his long-haul services. The Board decided to abandon attempts to see further institutions and went instead to a meeting with Richard at his house in Holland Park at 7.00 am the next morning. We took with us representatives of Barings and County NatWest and during a working breakfast we ran him through the broad scheme of things and discovered that he was very enthusiastic. We agreed to go into detailed discussions and we then adjourned.

That meeting was to become somewhat well remembered by all present, for upon leaving, it was discovered that every car we had parked outside Richard's house had been wheel-clamped!

After some delay we went for meetings with Richard's legal advisers and finally agreed the structure of a press announcement to be put out late that night, which indicated that we were in dialogue. As a result we asked the Stock Exchange to suspend the shares from start of business on 28 September.

We worked all over that weekend and in the next week Richard had his team start the process of detailed operating and financial assessment. It all seemed to be going very well.

The broad concept of the deal was that Richard was going to take a £10 million investment in Davies & Newman plc, but he would have had that on the same terms as all the other shareholders in the Company. Richard would then have taken a management contract to participate in the airline as a subsidiary of D&N, but he would not have had an equity control, or ownership of the airline as such. In our view he would have had around 15-16% of the Company, so he would not be in a dominant position. Richard would have contributed the brand of Virgin and Dan-Air would be re-named 'Virgin European' and there would be total co-ordination of marketing and sales strategy with Virgin Atlantic. Richard would then have got a share of

...that brought forth a massive increase in sales performace and revenue that was sustained to the final days.

SO NEAR DISASTER...

the profit as royalty for use of the brand name.
 With Richard's equity contribution this would have given us a sufficient head of steam to get the £35 million or more new captial. The concept was working and I was now feeling very optimistic. Things started to unravel at a meeting between myself and Richard at 4 o'clock on 5 October at Holland Park. We needed to sign the Heads of Agreement in order to go onto the next stage of the proceedings. This Richard could not do, for he had opposition from some of his colleagues in Virgin Atlantic and still needed to talk to others. We therefore adjourned the meeting until 9 October. Which we duly did and Richard indicated that he could not carry a consensus of his management for the scheme and as a result it looked as if it would have to be called off".

The share price of Davies & Newman PLC, the airline's parent Company had been falling since early in the year, when they peaked at 101p. By late July the *'Daily Express'* was reporting that **'Dan-Air shares nosedive'**, by a third to 24p that week.

David James vehemently repudiated some of the wilder examples of Press speculation, in particular any formal threat from the CAA concerning licences, a poor engineering record and its effect on time-keeping.

The *'Daily Telegraph'* on the 28 and 29 September reported that attempts were being made to raise £40 million through a rights issue and talks with Branson, who was prepared to invest £10 million if the City put in the rest. In return Branson wanted Dan-Air's profitable routes to be branded with the Virgin name under the possible title of Virgin European Airways to act as feeders for his long-haul flights. This announcement prompted much speculation and hilarity amongst the staff; were they going to become 'born-again Virgins'?

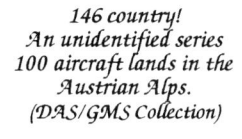

*146 country!
An unidentified series 100 aircraft lands in the Austrian Alps.
(DAS/GMS Collection)*

DAN-AIR SERVICES

Almost all the other daily newspapers reported the same story in one form or another, the *'Daily Express'* closing their piece on the 28th with the words that 'James yesterday refuted suggestions that there would be immediate job losses. *"Flying operations continue as normal"* he said'.

Speculation surrounded the airline, but David James repudiated stories that the airline had been given a deadline to find finance by either the Civil Aviation Authority or the Banks. *"At the end of the summer season, cash balances are between a £4m surplus and £3m debt"* he said. *"This is not a today problem, but a six months out problem"*. David James had tested City feelings to a rights issue during the week previously, so that Dan-Air could meet its spring peak borrowing period for 1993. Talks had been going on with Richard Branson since late August but, after making good initial progress, it became clear that by 12 October no deal was possible with Branson.

By now, the press were having a field day - not a day went past when an item of speculation failed to appear in the financial pages. This or that deal was close to conclusion and an announcement was expected 'to-day or early tomorrow', so-called sources 'close to Dan-Air' (whatever that meant) were quoted as saying.

In truth, when the media reportage was studied in detail, there was very little hard information. Even within Dan-Air, no-one had any firm information as to what was going on. Whatever talks were being held, their location and their contents were top-secret.

By 20 October the *'Daily Express'* was reporting **'BA plots a course for Dan-Air routes'** and the *'Daily Telegraph'* recorded **'BA's Dan-Air rescue move'**, whilst the next day *'The Guardian'* had **'BA close to Dan-Air deal'.**

The consensus of these reports was that British Airways had been involved in eleventh-hour talks to save the troubled operator. *'BA...'* reported the *'Express'* '*...headed by Chairman Lord King is anxious to get back routes to places such as Rome, Paris, Brussels, Amsterdam, Athens and Nice from Gatwick, which it was forced to give up in the take-over of British Caledonian'*. The report also hinted at the political side to events, when it went on to state that *'The Government, eager to avoid another announcement of mass redundancies, is believed to have given the move its tacit blessing and indicated that there will be no reference to the Monopolies and Mergers Commission'*.

The *'Telegraph'* revealed that BA was in talks with David James during September, but they broke up and James turned to Branson for help. According to the *'Guardian'* 'BA, which had never acknowledged involvement in talks to take over Dan-Air, would not comment yesterday on news that it was close to clinching a deal'.

SO NEAR DISASTER...

More behind the scenes details from David James:

" On Monday 12 October we started a dialogue with British Airways. They had first approached us in November 1991 at an Aviation Club lunch. BA had considerable problems at Gatwick, for they were, at that time, unable to operate their short-haul operation from there cost-effectively; they recognised that we had a lower cost-structure and there were possibilities for considerable rationalisation by putting our networks together. There might be scope for both of us to do each other a load of good.

We went into a dialogue right through until February that showed considerable promise. At that point BA held off until the arrival of John Olsen in April and things looked very promising for a while.

The idea, again, was an interesting one; British Airways would regard Dan-Air as a charter airline and we would be chartered to fly their scheduled network out of Gatwick. They would pay us the direct operating costs per flight plus a profit margin for doing so, and be responsible for all sales, marketing and booking. The question was: could we fly and operate at a cost structure that would integrate those networks, alleviate their losses and still provide BA with the essential interlining that they needed? Everyone thought so, for, as of the first quarter of 1992 Dan-Air's and British Airways' timetables listed each others flightcodes. In the end BA walked away from the discussions when they concluded that they did not believe that we had the financial resources to survive independently to do the deal.

However, the talks with Richard Branson failed, so I went back to them and said *'Look, I'm running out of road, I have nowhere to go at the moment and we're going to fail very swiftly indeed unless we have some help. I have no more than two weeks of survival left'*.

I met with them, and through 13 - 15 October we took half a dozen conference rooms at the Gatwick Hilton Hotel. We had their financial team and our financial team there, and we just took the books from Horley and sat them down to do an analysis - solidly through the night and day for three days - at the end of which BA came up with the view that it really was not on.

I had a tremendous battle to keep Lloyds Bank in place during these talks, for they wanted to pull out at that time. They were very worried. The bank had put Touche Ross in, who had done an assessment during late August, early September of the maximum value they would be able to realise on the assets in the event of our receivership and distress break-up. The had come to the conclusion that the value was a maximum of £29 million and that as we were always going to have £9 million of bonding outstanding (which they were exposed for) the maximum they would let our borrowings rise to was £20 million. The last day on which that figure was sustainable was going to be 29 October. Thus the banks had given me an absolute

*Boeing 727 'VT
(DAS/GMS Collection)*

deadline to have a solution in place by 29 October or we would go into receivership.

Finally, we found a way of doing the deal with BA - we had one, incredible non-stop face-to-face 26 hour meeting I remember - but we got to the point where there was a deal that was just about do-able. This was conditional upon all the paperwork being in place, to be able to announce it by the deadline dictated by our borrowings. Even then, on the night of 22 October we had actually had a receiver standing by to go in. We had worked out a flight plan for Friday 23 October whereby we had arranged for all the aircraft that were overnighted away (about six) to take-off as the receiver was timed to go in at 6.10 am London time - we would have grounded every flight at Gatwick that morning.

We finally made the decision to go ahead with the deal (although we did not sign until 11 am) and called off the receiver at 5.20 am, so we only had 40 minutes to spare!"

Whilst all this was going on, for many of the airlines staff - and their customers - it was very much 'business as usual'. If anything was going to happen it was outside their control - all they could do was give their normal high-as-possible service to the customers and hope for the best!

Then, on 23 October, readers of *'The Times'* business section awoke to a detailed report declaring **'Talks on Dan-Air rescue by BA hang in the balance'.** The bankers and lawyers it seems, were locked in talks arguing over the fine print. The piece said that in what was supposed to be a final meeting on Thursday afternoon, doubts were raised by Dan-Air's bankers, who feared that full repayment of their investment would not be guaranteed, and that Dan-Air would not be able to recover their losses.

It looked as if news was imminent...

DAN-AIR SERVICES

CHAPTER TWELVE
The news breaks...

The time was mid-day on 23 October 1992. The event was the release to every Dan-Air employee of a copy of *"Chairman's Newsletter No.16"* from Davies & Newman Holdings PLC. The period of speculation was over - now everyone had an indication of their fate.

The statement, consisting of five pages over the signature of David James, explained what had and what was about to happen...

"By the time you receive this Newsletter, everybody will have heard the news of the deal with British Airways PLC announced today.

Under the terms of this transaction, and provided a number of essential pre-conditions can be fulfilled, Davies & Newman Holdings PLC has agreed to sell the whole of its assets and undertaking to BA. This means that Dan-Air Services Limited will become a wholly owned subsidiary of BA. In return BA have undertaken to settle all the liabilities of Davies & Newman.

Given the volume of press comment over the last couple of weeks, this news may hardly come as a surprise. Even so, I know that you will all be extremely concerned to know what the transaction means to the airline as a whole and for each of you personally.

You will all feel that this outcome is a major disappointment when compared to the high hopes we had after the successful fund raising almost exactly a year ago. Nevertheless, the Board and I are convinced that the only alternative to this transaction with BA would be the collapse of the Company and the immediate termination of all operations with very much worse consequences for everybody concerned than we may now anticipate.

The Board of Davies & Newman is initiating a programme to reduce the scale of Dan-Air's operations which involves the following steps to be implemented before completion.
* the cessation of Dan-Air's charter operations;
* a reduction in the number of destinations served by Dan-Air's scheduled network from 28 to 12;
* disposal of 26 of the 38 aircraft operated by Dan-Air during the summer of 1992, leaving a reduced fleet comprising of only 737-400s to operate a smaller scheduled network; and
* a major reduction in the number of staff employed in the business to the level required by the on-going operation. We currently have some 2,000 permanent staff and, of these, between 400 and 600 will be offered continued employment. Of the balance, those aged 50 and over will be given the opportunity for early retirement. The remainder will sadly be made redundant but on the terms to which they are contractually entitled. This, as I have commented earlier, may be considerably better than could have resulted in a receivership".

The newsletter went on to explain that :

"However disappointing the present situation may be, it would be a very great deal worse than without the BA deal".

DAN-AIR SERVICES

To say that everyone was devastated by the news is an understatement. True, a deal of some kind (possibly with BA) had been expected. As we have already seen, there had been much speculation, especially on the business pages of the papers, and it had became clear that some kind of deal was in the offing that involved British Airways. It was not the news of the deal that shocked everyone, but the scale of the cut-backs.

According to the Newsletter, the Company was, to all intents and purposes, bankrupt. In many respects this was contrary to the information that the staff had been receiving. During the previous few weeks everyone had been kept informed as far as the rules for suspension of shares on the Stock Exchange allowed; even the outstations received regular telexes from the Chief Executive. The information that had been promulgated was, perhaps understandably, reasonably upbeat considering its source.

The deal meant that at least some of the aircraft would still be flying, passenger tickets would be guaranteed and the banks certain of getting their investments back, but many members of staff felt that they had been ill-prepared for the news. "*If only we had had some idea...it would have helped somehow*" was heard more than once over the next few days.

What had happened?

The Chairman's Newsletter attempted to explain. At the time of the last fund-raising by Davies & Newman in the autumn of 1991 there had been widespread expectations of a modest economic upturn during 1992 with an assumed growth in Gross Domestic Product (GDP) of 2.6% which, by a formula used in the airline industry, implied traffic growth of 5.2%. In the event, instead of a recovery, there was a decline in the UK GDP and the economic prospects for the future appeared as bleak as at any time during the recession.

This economic background had a severe effect on the finances of Dan-Air. In the six months to 30 June 1992, the Davies & Newman Group incurred an estimated operating loss of about £24M.

Once it became clear that the pattern of trading for 1992 was likely to fall far short of expectations, the entire operation was reconsidered. On the scheduled side, a need was identified to reduce capacity by cutting back on flight frequency and closing routes where inadequate loads were likely to occur. Several of the new routes failed to meet expected targets so, as a result, the services to Stockholm and Barcelona were closed. For the first five months of 1992 charter performance was only marginally below expectations, but this showed a progressive worsening from June onwards.

When both sets of figures were placed together, the shortfall in generated revenue meant that Dan-Air began to fall considerably

THE NEWS BREAKS

below the level necessary to provide the resources to trade through the on-coming winter season.

At the time of the take-over, the level of Group borrowings was fully covered by the security available to the Company's bankers. However, without further new capital Dan-Air's normal seasonal pattern of operations, coupled with the continuing adverse economic circumstances, would have meant that bank indebtedness would have risen to a far high level over the period to April 1993. It was likely then that this would have greatly exceeded the facilities available from the Group's bankers.

In the event, the Board concluded that, given the Company's financial position, it would not have been possible to implement a new fund-raising to enable the Company to continue in operation on a independent basis.

A number of discussions had taken place (according to David James in one interview, twenty in all) with other airlines and third parties. All failed to find a solution to the Company's difficulties.

The only alternative appeared to be receivership. This would have resulted in the forced sale of the Company's assets which, together with the additional liabilities that would have crystalised would have almost certainly eliminated the Group's entire nett assets. This would have resulted in only a small portion of the total sums owing being paid to the Group's creditors and total loss of employment for all the Group's staff.

Thus it was thought that the transaction with British Airways would at least achieve some benefits (although provide nothing for shareholders) and a drastic, but not total loss of employment.

British Airways' side of events...

News of the deal was released to the waiting world in two documents. Firstly, there was the Chairman's Newsletter from Dan-Air; then there was a three-page News Release under the

737-400 G-TREN seen at Newcastle in the final Dan-Air colour scheme. The change is minor - the word 'London' has been dropped from the fuselage titles. The entire fleet was scheduled to undergo this change at the time of the take-over, but only a few were completed.
(DAS/GMS Collection)

DAN-AIR SERVICES

heading **'Dan-Air's Key Gatwick Routes saved'** issued by the Press Office of British Airways.

This second document, broken up into short paragraphs and with short, sharp quotations attributable to Sir Colin Marshall, Deputy Chairman and Chief Executive of British Airways, provided a valuable insight into the other side of the story.

The early part of the Release dealt with the conditional deal, the fact that BA would pay a nominal £1 for Davies & Newman's remaining assets and the routes it was continuing to serve. Then came details of the plans...

> "This move is the first stage in British Airways' strategy to create a new, low-cost airline dedicated to the continuation and further development of a major network of short-haul services at Gatwick.
>
> The subsidiary will operate the Dan-Air scheduled Gatwick routes using a fleet of 12 modern Boeing 737 aircraft. It will also operate the existing ten British Airways Gatwick-based short haul routes, with its fleet of 20 Boeing 737s.
>
> The plan is progressively to standardise this fleet on the most modern Series 400 version of the twin-jet, and consolidate operations at Gatwick's North Terminal".

The quotations from Sir Colin were laid out so that reporters and news editors could select the one that fitted the tone and slant of the piece they were writing...

> "The opportunity to save Dan-Air's scheduled network coincides with our own objective of reducing significantly the costs of our shorthaul operations at Gatwick."
>
> "We have made good progress towards this goal in the last 18 months or so - we now have a means of achieving it, safeguarding our shorthaul operations at the airport in the long term".
>
> "We regret the circumstances which made the future so uncertain for Dan-Air. It has been bad news for a good airline".
>
> "The implications of its possible demise for Gatwick, as a viable scheduled service hub, has given us great cause for concern in recent weeks".
>
> "We are, therefore, pleased to have reached this agreement with Davies & Newman after being approached by them again".
>
> "The arrangements we have concluded enable us to secure Gatwick's vital shorthaul links and save the jobs of several hundred people, by operating the amalgamated network on a cost-efficient, competitive and profitable basis, producing profits for future development".
>
> "Tough recessionary times produce tough solutions. Our vital international transport links need to be preserved and expanded above all".

The release concluded with some facts and figures.

> "In the year to December 3 1991, Davies & Newman Holdings PLC reported a pre-tax loss of £35.4 million on a turn-over of £325.3 million, In the six months to June 30 1992, it incurred an estimated operating loss of £24 million.
>
> British Airways will fund the costs of the arrangements from cash reserves..."

THE NEWS BREAKS

However, one other source had revealed in advance some interesting snippets of information about BA's plans - an item that appeared in the British Airways News, dated 22 October! Under the headline **'What the Dan-Air deal means to you'** the team working on the Gatwick link-up outlined the implications for British Airways staff at the airport. It explained that BA was setting up a new shorthaul airline at Gatwick using, where appropriate, aircraft, pilots and cabin crews leased from British Airways.

Reaction of the Media

The news was released of the take-over at mid-day, just in time to catch ITN's 12.30 bulletin. This was followed by the BBC's One O'Clock News, where it was the second lead item. *"British Airways is taking over Dan-Air. About fifteen hundred jobs will go"* the newsreader's voice intoned over shots of a Dan-Air 737 taxying out to the runway at Gatwick. After outlining the terms, Christopher Wain, their Aviation Correspondent, explained that meetings had taken place overnight between the two airlines' lawyers at Speedbird House, British Airways' Heathrow Headquarters. He went on to state that negotiations had taken place about a rescue package previously, but had broken down, as had the talks with Virgin Atlantic. Much of the rest of the item had been taken from the Chairman's Statement, although it concluded with information that a number of airlines - Britannia, Air UK and British Midland - were already voicing concern about the deal. Indeed, Sir Michael Bishop, British Midland's Chairman was interviewed in the report, during which he stated that his airline had already had meetings with the EC Commission in Brussels and had made representations in writing to the Office of Fair Trading and the Department of Transport. He said they would be seeking vigorous scrutiny of any deal between Dan-Air and British Airways.

From the news item it appeared that the Unions were relieved; it seemed that they had feared many more job losses. George Ryde, a Transport & General Workers Union official, said over the telephone to a reporter;

> "It certainly could have been worse. We need a lot more information to see the details of this package before we would want to say that this is the best thing that could have happened. But without a doubt, it could have been much, much worse than it is".

Virgin Atlantic's Richard Branson was also featured. Virgin wanted to compete with BA on European routes, which as the report stated *"by implication meant that British Airways had to surrender flight times, known as 'slots'"*. In an interview Richard Branson said *"...we must be able to fly alongside British Airways to places like Paris, Brussels and other cities in America. I think that as long as we can actually get those necessary slots, we have no real objection to BA getting bigger"*.

DAN-AIR SERVICES

Christopher Wain concluded his report by saying *"For the Government, a BA rescue has the advantage of saving at least some jobs at Dan-Air. But any link-up, however loose, between Britain's two biggest airlines will make life harder for the rest. And even if it escapes reference to the MMC (Monopolies and Mergers Commission) it's still possible that the deal will be taken to the European Commission"*.

There was worse to follow. Fleet Street went mad. There were camera crews all around Gatwick, reporters trying to talk to the staff and photographers jostling to take pictures of weeping girls who were still striving valiantly to do their jobs. Locally, the author received a frantic phone call from ITN's Picture Library who wanted a couple of photographs about the early days from his personal collection to use on 'News at Ten'. Before agreeing to send them, approval was sought and gained from Dan-Air's Public Relations Department. That in itself was no small task, for all lines into Newman House were constantly engaged.

By the early evening news the story had become the lead item. John Suchet, reading the 5.40 ITN headlines, said that *"For the cost of a bus-fare, British Airways this afternoon bought an airline. It cost BA just a pound to take over their ailing rival Dan-Air. BA will take on around half of Dan-Air's scheduled services but Dan-Air's charters will be closed down. BA will also take on Dan-Air's debts of several million pounds and between 1400 and 1600 Dan-Air staff will lose their jobs"*.

Greg Wood, ITN's Aviation Correspondent, had filed a report that reflected what had been happening since the news broke at lunchtime. His report quoted David James as saying that *'the Company had been a basket case when he was appointed in 1990 and that rescuing it had become a yearly event. What had eventually sunk it was a sharp decline in the number of charter passengers'*. Most of the rest of the items were again taken from the Newsletter.

Since the news broke reporters had been doing more work, for now there were comments attributed to British Airways, mainly taken from the News Release issued by BA's Press Office. Wood continued *"BA said it would cost around £35 million to pay off Dan-Air's debts and that the bid wouldn't go ahead if the UK competition authorities opposed it. The 400 Dan-Air staff who keep their jobs will be offered lower terms, but BA was unapologetic..."* Much of the early part of the report was broadcast over images of Dan-Air at Gatwick, but now the television picture changed initially to show a British Airways team, including their Chairman, Lord King of Wartnaby and Sir Colin Marshall, Deputy Chairman and Chief Executive, arriving at an office building, possibly for a press conference. and then Lord King, Sir Colin and

THE NEWS BREAKS

David James at the conference before showing a close up of Lord King, apparently responding to a question. *"The deal that they have, at midnight tonight young man, is no job, no deal. We're not taking advantage of that, what we're saying is there will be new, er, employment offered in the new company"*.

David James was then interviewed. *"I believe it was highly likely that Dan-Air would have been forced to cease flying this morning without the assurance that this transaction was available to the Company"*.

Greg Wood moved on to the wider aspects.... *"BA will now inherit Dan-Air's take-off and landing slots at Gatwick Airport. But it is already facing competition from other airlines. Virgin today applied for two of Dan-Air slots that BA wanted to keep..."*.

There were interviews with Richard Branson and Sir Michael Bishop. Wood then 'wrapped up' his report... *"Dan-Air's charter operations will be closed before the BA take-over (then) the Dan-Air name, first seen forty years ago, will disappear for good"*.

The report concluded with a short piece 'live' from the spectators terrace at Gatwick, where Joan Thirkettle told travellers holding tickets that she had spoken to both Dan-Air and British Airways and had been assured that *"all bookings would be honoured, all tickets were valid"* and to check with their travel agents.

Less than 20 minutes later more or less the same story was repeated on the BBC 6 o'clock news, obviously built on what they had already broadcast at 1 pm.

More of the press conference featuring Lord King was shown, during which he revealed more of what was happening. *"The Company is bust"* he said, *"It's closing down, or will close down. We are picking it up. We are re-structuring the Company. We're offering jobs to people that will be out of jobs. Now, whether the pay is, er, competitive with what they were having, or what they were getting or not, I can't tell you. Those rates have to be agreed, but they're good jobs, and there's a tremendous opportunity here to build a new and strong and powerful Gatwick"*.

David James also revealed more of what had been going on in recent months. *"... had gone to the limit of the Company's ability to survive without an alliance. Throughout the last four months we have realised we were under progressive pressure and we had sought an alliance with one of the leading airlines of the world. We've actually had dialogues with some twenty such operators and failed to come up with any meaningful discussion"*.

The report closed with further aspects. The Civil Aviation Authority was quoted as saying *"...it's raised serious issues of competition"*. Other airlines, it said, felt the same way and showed further interviews with Branson and Bishop.

The BBC's nine o'clock news more or less repeated the same story. The ITN's New s at Ten also contained much of it's 5.40 broadcast, but also went into the story in scanty depth, detailing some of British Airways' recent acquisitions and showing briefly the two photographs from the author's collection obtained earlier that day. The broadcast media then promptly lost interest.

Saturday saw the turn of the newspapers. Many broadly reported the facts as they had them. **'BA buys ailing Dan-Air for £1'** said *'The Times'* whilst the Air Correspondent reported in a semi-historical piece why **'Frantic maydays to City could not keep Dan-Air aloft'**. This stated that the root cause of the difficulties appeared manifold and lay in the bones of the Company itself. It had an old fleet of aircraft that were expensive to operate and maintain; it had no tour operator to guarantee a flow of passengers to its charter services and, in operating from Gatwick rather than Heathrow, could not attract sufficient business customers. *'The final blow...'* the piece said *'..came when rumours began circulating around the industry that it was in serious trouble and might not survive the winter'*.

The worst headline - to a piece containing a number of blatant inaccuracies - appeared on page five of Saturday's *'Daily Express'*. **'1,600 jobs are lost as Dan Dare is grounded'.** The airline was not grounded, and the reporter even credited Sir Michael Bishop as being Chairman of Britannia Airways, not British Midland!

The aftermath.
Reports in the press soon dried up. On Saturday the author visited Gatwick. The red, white and blue jets were still taking to the air, the bright blue uniformed airline staff and the staff of Gatwick Handling were still manning the ticket and check-in desks and the passengers were still checking in.

Surprisingly, the mood, considering what had happened, was remarkably positive. The staff may not have known what the future held, but all were determined to continue to give the passengers the service they deserved. As John Aston, one Duty Manager, said *"I intend to work right up to my last day, giving the best I can"*. Interestingly, considering what had happened in the previous 48 hours, the travelling public were voting with their feet, for Dan-Air's scheduled services were almost all over-booked!

The next two weeks were to be surreal times. It was 'business as usual' in some departments, a rapid close-down in others. All members of staff received redundancy notices that came into effect on 2359 hrs on Friday 6 November. Those that were going to be re-employed were to be notified later and placed on new contracts.

The employees may have been battered and bruised by the media attention, lost their livelihoods and were immensely

THE NEWS BREAKS

*1500 hrs, Saturday 24th October 1992. The news has finally sank in, but the staff manning the Ticket Sales desk at Gatwick are still providing the customers with what they require! In a few days, the desk was covered with cards from customers expressing sympathy.
(DAS/GMS Collection)*

saddened by the events, but all were determined to go out with their heads held high, proud to be part of Dan to the last.

Of course, there were moments of weakness. Some Cabin Staff could not complete the *'Thank you for flying Dan-Air and we hope you travel with us again'* announcements at the end of the flight - or if they did, it was while blinking back the tears and choking back the sobs. One security guard was heard to say *"Dan cannot go, you're part of the furniture"*. Even passengers were seen with tears in their eyes - it was like losing an old friend.

The national newspapers may have soon forgotten the story, but the next week's travel trade newspapers had plenty to say. Edition 1138 of *'Travel Weekly'* was a good example. Headings such as **"BA Gatwick staff face salary cut"**, **"Angry rivals predict loss of competition"**, **"Little left for charter carriers"**, **"BA deal sparks rush for scheduled routes"** and **"Death of a charter legend"**. However, the 'Comment' section of that paper probably best summed it all up.

"...Without an in-house tour operator it was severely exposed to the downturn in the charter market in the late 1980s leading to its first loss in 1989. Adverse market conditions made it impossible for Dan-Air to build up its scheduled services fast enough to compensate.

In the end the deal with BA was the best Dan-Air could hope for, but the move nevertheless provoked angry cries of 'not fair'.

But those who are now complaining, including British and foreign airlines, had plenty of opportunity to block BA's move. Dan-Air chairman David James approached 20 institutions for help, but none were prepared to take the risk of investing in the airline.

DAN-AIR SERVICES

So all of those who are now complaining, having failed to put up, should shut up".

The deal between Davies & Newman and British Airways had left the shareholders with nothing. A letter with attached Press Release from Cardew & Co was sent to all shareholders explaining what was happening. They were not given the chance to vote on the proposals because, David James said, there was a need to move quickly and the deal represented the only alternative to receivership. He is reported to have said "...*that by the time a poll could be arranged the Company would have exceeded all borrowing limits and be insolvent. Special consent had therefore been granted by the Stock Exchange to waive the vote*".

The statement concluded with two paragraphs on the winding up of Davies & Newman.

"The Board of Davies & Newman intends to recommend that, following completion of the proposed transaction with BA and the discharge in full of its liabilities, the Company be wound up.

A circular giving details of the proposed transaction and containing notice of an Extraordinary General Meeting of shareholders to consider the members' voluntary winding up of Davies & Newman, will be posted shortly".

'The *Daily Telegraph*' reported that one senior institutional fund manager whose company had a substantial exposure to Davies & Newman insisted on anonymity, but said "*We are not very happy, but what can we do? These things unfortunately happen.*

"*We took the shares on recovery hopes and that did not happen. Frankly there is a not a lot I can tell you about this, but it has been a disaster. The bad news is there is virtually nothing left for the shareholders*".

Another anonymous City source was quoted in the same article. "*It's probably good news for the bankers, but there's no prospect that the shares have any value whatsoever*".

On 2 November the Government, in the form of Michael Heseltine, Trade and Industry Secretary, and Sir Bryan Carsberg, Director General of the Office of Fair Trading, gave the take-over the the go-ahead, despite other carriers' worries about unfair competition. Heseltine and Carsberg acknowledged that the rescue raised competition issues, because BA would have a bigger stake at Gatwick as well as Heathrow, but worries about damage to the former, if Dan-Air was allowed to collapse, had led Heseltine to give the all clear.

Several independent airlines were continuing to press Sir Leon Brittan, the EC Competition Commissioner, to investigate the competition aspects to the deal, although Sir Leon had already said that the take-over was not big enough to qualify for EC study. It was all over bar completion of the paperwork.

A number of farewell parties were held. It was like attending a

THE NEWS BREAKS

Scheduled Services Timetable

The final scheduled services timetable - for the summer of 1992. (29 March 24 Oct.)

two week long 'wake'. Possibly the biggest - in the Desoutter Suite of the Forte Crest Hotel, adjacent to Gatwick's North Terminal on Wednesday 4 November - was primarily organised by Lisa Maunders of Dan-Air Passenger Services and others from Gatwick Handling, for they had been affected as well as the airline staff. Around five hundred had been expected that evening, but eventually many times that number were packed like sardines into the ballroom, bar - even the cloakrooms!

There were staff from all over Europe, people from Gatwick Handling, engineers from FLS... There were hugs, tears, exchanged addresses... and much sadness. But not defeat! They had lasted longer than everyone else, they had given it their best shot. None present felt that there was anything to be ashamed about. Mike Street, the Managing Director designate of the new British Airways operation, acknowledged the staff's professional and correct behaviour right to the end at a meeting attended by Mike Forsyth - an accolade that was repeated across the industry about how everyone within Dan-Air handled themselves and their business during the last few days.

Bryn Wayt, Deputy Company Flight Safety Officer, sent an open letter to the staff and newspapers:

"In the normal run of high density flying in and around Europe, the stress and strain, and mental pressures of the job is well documented.

When your company is under threat of foreclosure and your livelihood and way of life is about to change out of all recognition, then the job becomes immeasurably difficult.

In the last two weeks of Dan-Air's existence - that lasted nearly 40 years - over 300 pilots and flight engineers flew knowing their career with Dan-Air was coming to an end as well as their airline.

I want the aviation world to know that those flight deck crew members did not let their standards drop one notch, and not a scratch was put on an aircraft or a passenger - despite these professionals operating under the intolerable strain. Tears were shed before and after flights without shame.

Much is written about flight safety. If examples are to be made about how it works, look no further than those who served Dan-Air so well, and then draw the curtains.

To my friends and colleagues who flew so well, happily and safely right to the end and are now missing from the crew room - I salute you!

Aircraft and staff began to drift away, the aircraft back to the leasors -the 727s via Lasham - and the people to their homes with three months' pay in lieu of notice in their bank accounts. One pilot on the evening flight from Heathrow to Inverness reported to the passengers over the Cabin PA that *"the weather is so clear tonight, I can see right through the doors of the Unemployment Office!"*

The charter flights quickly changed to other carriers and on the scheduled service routes that BA were not keeping other airlines took over, but not without difficulties. For example, the last DA scheduled service from Newcastle was DA107 (flown by 1-11 G-BDAT) which left for Gatwick on 2 November. After that the

DAN-AIR SERVICES

THE END OF AN ERA
DAN-AIR SERVICES LAST COMMERCIAL FLIGHTS.

Last BAe 146 flight -	Flight Number DA081. Aircraft was G-BUHB, a BAe 146-300. Flight was from Aberdeen to Gatwick. Arrived Gatwick 0857hrs Friday 6 November 1992 in the hands of Captain Hutchinson. On board were 46 passengers.
Last 727 flight -	Flight Number DA758/9. Aircraft was G-BNNI, a 727-200. Flight was from Gatwick to Fornebu (Oslo), and return. Arrived Gatwick 2348hrs Sunday 1 November 1992 in the hands of Captain Lenton. On board were 32 passengers.
Last 1-11 flight -	Flight Number DA910. Aircraft was G-BCWA, a 1-11-500. Flight was from Toulouse to Gatwick. Arrived Gatwick 0937hrs on Friday 6 November 1992 in the hands of Captain Threlfall. On board were 34 passengers.
Last 737A flight -	Flight Number DA151. Aircraft was G-WGEL, a 737-200. Flight was from Inverness to London Heathrow. Arrived Heathrow 0815 hrs on Friday 6 November 1992 in the hands of Captain Cherry. On board were 40 passengers.
Last LGW arrival -	Flight Number DA948. Aircraft was G-BSNV, a 737-400. Flight was from Montpellier. Arrived 2155hrs on Sunday 8 November in the hands of Captain Perry. On board were 104 passengers.
Last Charter flight -	Flight Number DA2558/9. Aircraft was G-BVNN, a 737-400. Flight was Gatwick to Gran Canaria and return. Arrived Gatwick 0819 on Monday 2 November 1992 in the hands of Captain Smyth. On board were 156 passengers
Last Dan-Air flight -	Flight Number DA689. Aircraft was G-BNNK, a 737-400. Flight was from Gatwick to Madrid. Arrived Madrid 2232hrs on Sunday 8 November in the hands of Captain Dogherty. On board were 56+1 passengers.

service was operated by Cityflyer Express using ATR-42s. Many of these flights have since been over-booked, due to the reduced capacity of the ATR-42 when compared with the BAC 1-11.

On what was supposed to be the final day, a number of British Airways aircraft and crews began to appear on Dan-Air routes. For instance, G-WGEL, the night-stopping 737-200 that overnighted at Inverness, operated the early morning service down to Heathrow with its normal crew, but for the rest of the day the service was flown by a BA machine. Those that were left working were completing the arrangements for the final hand over and waiting for 2359hrs.

The Final Day of Dan Air.
A number of 'last' flights were duly recorded on 6 November, but as the gangs who changed all the signs on Line 5 and the ticket desk at Gatwick from Dan-Air to British Airways were to find out, this was a false sunset! In the middle of the night Simon Leighton, Duty

Seen parked out of service in front of the FLS hangar at Lasham, on 7 November 1992 are Boeing 727-200s G-BPNY and G-BPNS (Robin A. Walker)

THE NEWS BREAKS

The final Dan-Air 'In Flight' magazine for Summer/Autumn 1992.

Manager Operations, urgently requested that all the Dan-Air signs, displays and advertising were quickly put back into place and Dan-Air rose again for another two days of existence.

The reason why Dan-Air arose phoenix-like for two more days borders on farcical, as David James explains:

"As far as I was concerned the whole deal was going through in an orderly manner, so I set off on 4 November for another board meeting in the USA. I did a day's work in New York on the 5th, expecting the whole agreement to be signed in good order in London on Friday 6 November. What then happened was just farce.

First of all, in order for the deal to be completed, every aircraft that was due to be sold on the charter side had to be sold by close of business on 6 November. In fact, everything went smoothly but for two aircraft for which the team in the UK could not obtain security release, and there was an argument over some minority financial interest who would not accept the terms they were being offered.

Finally, everything was ready to go, except that they could not find any way of getting the security cleared by one of the American banks that was based in California. The London representative of the bank (who had not been told he was needed) had gone off to a wedding somewhere in Yorkshire and no-one could find him and there was nobody else with the power of authority to sign it.

So we got onto the banks head office in Los Angeles only to discover that the only two directors with power of authority to sign it were part of a management think-tank which was being held - of all places - under canvas in the Sierra Nevada desert!

We ended up getting a portable Fax machine airlifted out to them by helicopter so as to be able to get a signature on the security clearance in the early hours of Saturday 7th and then Fax it back to the UK so that there was the authority to clear and close the whole deal!

In order to do it all, I was flown back from the USA aboard Concorde, where I experienced what is possibly the ultimate ego trip - I had established for me a radio link from the cockpit of the aircraft so that I could, if necessary, maintain constant contact with the conference in London. We finally completed the whole thing at exactly 19 minutes past 12 on Sunday 8 November - that was the historic moment.

With the deal finally signed, operational control could move to British Airways at 2359hrs that night. All the signs were changed once again and the aircraft remaining in service had 'British

\multicolumn{6}{	l	}{LAST FLIGHTS/PARKING POSITION FOR STORED OR DISPOSED A/C}
Date	Type	Regn	Depart	Arrive	Pilot
28/10/92	727-200	G-BHVT	Gatwick @ 0830hrs	Lasham @ 0918hrs	Capt. Marson
02/11/92	727-200	G-BMLP	Manchester @ 1300hrs	Lasham @ 1337hrs	Capt. Duff
02/11/92	727-200	G-BPND	Manchester @ 1318hrs	Lasham @ 1358hrs	Capt. Budd
02/11/92	727-200	G-BNNI	Gatwick @ 1428hrs	Lasham @ 1453hrs	Capt. Newton
02/11092	727-200	G-BPNS	Gatwick @ 1505hrs	Lasham @ 1543hrs	Capt. Burgess
03/11/92	727-200	G-BPNY	Gatwick @ 1308hrs	Lasham @ 1349hrs	Capt. Kloos
06/11/92	146-300	G-NPNT	Gatwick @ 1144hrs	Filton @ 1227hrs	Capt. Buist
06/11/92	146-300	G-BPNU	Gatwick @ 1150hrs	Filton @ 1235hrs	Capt. Barker
06/11/92	146-300	G-BUHB	Gatwick @ 1219hrs	Hatfield @ 1259hrs	Capt. Reiss
06/11/92	146-300	G-BUHC	Gatwick @ 1233hrs	Filton @ 1316hrs	Capt. Hertzberg
02/11/92	1-11/500	G-BJMV	Gatwick @ 1118hrs	Bournemouth @ 1218hrs	Capt. Rooke
02/11/92	1-11/500	G-BJYM	Gatwick @ 1320hrs	Bournemouth @ 1400hrs	Capt.Threlfall
02/11/92	1-11/500	G-BJYL	Gatwick @ 1340hrs	Bournemouth @ 1424hrs	Capt. Barnett Higgins
03/11/92	1-11/500	G-AWWX	Gatwick @ 1210hrs	Bournemouth @ 1300hrs	Capt. Threlfall
03/11/92	1-11/500	G-BCXR	Gatwick @ 1234hrs	Bournemouth @ 1312hrs	Capt. Hutchings
05/11/92	1-11/500	G-BDAS	Gatwick @ 1950hrs	Southend @ 2025 hrs	Capt. Pickering
06/11/92	1-11/500	G-BEKA	Gatwick @ 0907hrs	Bournemouth @ 0952hrs	Capt. Keating
06/11/92	1-11/500	G-BDAT	Gatwick @ 0742hrs	Southend @ 0817 hrs	Capt. Berryman
06/11/92	1-11/500	G-BDAE	Heathrow @ 0954hrs	Bournemouth @ 1043hrs	Capt. Butler
06/11/92	1-11/500	G-AXYD	Gatwick @ 0959hrs	Bournemouth @ 1041hrs	Capt. Griffiths
06/11/92	1-11/500	G-BCWA	Gatwick @ 1030hrs	Bournemouth @ 1105hrs	Capt. Noble
02/11/92	B737/200	G-BMDF	Manchester @ 1548hrs	Gatwick @ 1638 hrs	Capt. Cook
05/11/92	B737/200	G-BLDE	Manchester @ 0651hrs	Gatwick @ 0745hrs	Capt. Mugridge
06/11/92	B737/200	G-BKNH	Manchester @ 0650hrs	Gatwick @ 0742hrs	Capt. Mugridge
06/11/92	B737/200	G-WGEL	Heathrow @ 0971hrs	Gatwick @ 0957hrs	Capt. Cherry

DAN-AIR SERVICES

Airways' stickers placed over the Dan-Air' titles, prior to repainting in the full British Airways livery.

The last ever Dan-Air flight.
Details of the first Dan-Air flight are sketchy. However, with the passage of 40 years and the benefits of modern technology, the methods of recording such details have improved and it is only right and proper that the fullest amount of information is recorded about Dan-Air's last flight.

The yardstick used to determine the 'last flight' was that it had to be a genuine Dan-Air aircraft (in full colours) with their own flight deck and cabin staff and Company call-sign/flight code.

According to SCOPE, the DAN-OPs computer, the honour of the airline's last flight fell to G-BNNK, a 737-400, at the hands of Captain Digby P. Dogherty and First Officer Mike Richards. The 56+1 passengers aboard were looked after by Cabin Staff No.1 Sarah Hoad and Annette Grace, Isabel Lafaia, Guy Patten and Hazel Robertson. In the hold was also 2317kg of freight. The flight, DA689 from Gatwick to Madrid, pushed back from the stand at 2021hrs, commenced taxi at 2022, was airborne at 2030. It landed at Madrid at 2222hrs and was on chocks at 2232. The crew ended their duty period at 2302hrs - it was the end of an era.

The last act of defiance - not everything wanted to join British Airways! 737-300 November Juliet still refuses to give up the Dan-Air name. Immediately after the deal with BA was concluded, engineers applied large "British Airways' stickers to obliterate the Dan-Air name. However, sometime during a flight from Paris to Gatwick, the rear part of the sticker came off, which resulted in this totally hybrid scheme. (Photo: Mike Forsyth)

CHAPTER THIRTEEN
A personal view...

Since the original conception of this book, it was always my intention that the final chapter was to be set aside for the Board of the Company to explain their vision of the future for Dan-Air. Instead I found myself writing their obiturary.

For ten glorious months during 1992 I had lived, eaten, slept and breathed Dan-Air Services. I had experienced seeing the beautiful wildness of the Shetlands from the flight-deck of a 748 on a glorious summer day as we approached that tiny strip of tarmac called Scatsta. I had drunk literally gallons of in-flight coffee in the soft glow of CRTs on the flight-deck of a 737-400 as the outside air temperature reached minus 59°C. during a night flight from Gatwick to Antalia in Turkey and back, the story of which, in the end, was never used. I had enjoyed the enthusiasm and dedication of Mo Perera, laughed with many at myself for some of my stupid questions, talked over mutual fears with Margaret Gray and her outstation staff whilst waiting to come home from Mahon and had cried with all of them on 23 October.

Now the dust has settled, the name has all but disappeared. But Dan-Air lives on in the minds of many. Possibly I had a unique viewpoint:- certainly many that worked within the Company felt that way. In the preceeding chapters I have tried to keep my own personal opinions out of the story so as to record as accurately as possible what 'The Spirit of Dan-Air' was all about. But, in the light of the monumentous events, there comes a point where my feelings whilst seeing it all have to be recorded.

What made Dan-Air what it was? There are many answers to that. The professionalism, the equipment and the service provided were all tangible aspects, but there was so much more. Everyone working in the Company made up one big happy family and the roots of that attitude go back many years to the man who everyone simply referred to as Mr Newman. This also applies, in a wider respect, to the Newman family.

In establishing Dan-Air and nurturing the airline to its peak, Fred Newman built a caring concern founded upon sound commercial principles; this caring nature of 'the Company' (allowing staff as much time off as possible when a family member was sick for example) was paid back many times over by a high degree of staff loyalty to both the airline and the Newman family. Throughout the Company and all the subsidiaries, there were many people who had stayed for most of their working lives, although they could have got better-paid jobs elsewhere. This devotion and enthusiasm was what 'The Spirit of Dan-Air' was all about.

DAN-AIR SERVICES

A question of image.
During the last few months of Dan-Air's operations, the so-called 'experts' within the press had a field day raking over what they saw as the root cause of the Company's ongoing difficulties. There was much comment about the airline's poor brand image, although many had the good grace to admit that this 'perceived image' was totally undeserved.

But from where had this 'poor brand image' been derived and who was seeing it? It is difficult to judge, although it has dogged Dan-Air and its employees for many years. In an attempt to understand this more fully we have to go back in time and look again at the early days.

During the 1950s and 60s the United Kingdom's social structure was much more stratified than today and in many ways this was reflected across into civil aviation. The state airlines saw themselves as primarily serving the middle and upper classes, whereas the charter outfits - well, they were for the workers and their packages in the sun! British European and British Overseas Airways Corporations (then both state owned) carefully cultivated and exploited a 'better than thou', almost snobbish attitude towards the independents. This 'lower class' appellation attached itself not just to Dan-Air but to all the charter airlines.

However, the charter airlines were all operating unsubsidised by the Government; they were subject to commercial pressures and market forces that the two state airlines never really had to consider. Money could not be wasted on obtaining expensive new equipment; the bottom line was profit and 'bums on seats' were vitally important for their very survival.

This led to the charter airlines obtaining aircraft at the lowest possible cost and then flying them to the maximum possible utilisation. This high usage, coupled with the undoubted higher rate of mechanical problems associated with large piston engines and comparatively primitive radio, navigational and other support equipment, all created a situation where flights were prone to

Culled from the Annual Reports to shareholders, the figures in the chart below demonstrate the performance of the airline from 1967...

	1967	1968	1969	1970	1971	1972	1973	1974	1975	1976	1977
Employees	538	651	842	1059	1435	1801	1973	2016	2164	2794	3020
Load factor (%)	75	83	86	82	84	84	81	78	82	78	82
Pax miles flown (Millions)	164	268	556	677	1191	1718	3384	3295	3923	4230	4977
Pax carried/sector (1000s)	247	275	565	746	1129	1741	2157	2193	2582	2846	3591
Turnover (thousands)	2,920	4,150	6,675	9,660	14,920	22,108	29,692	39,109	52,806	79,404	100,652
Airline Operation	n/a	n/a	n/a	n/a	14,123	21,346	28,756	36,176	51,780	78,382	99,433
Profit/(loss) for year	164	103	192	374	256	237	292	431	356	528	332
No. of aircraft	16	14	20	22	24	34	39	34	52	54	63

A PERSONAL VIEW...

departure delays. Obviously, this was not good for any airline's image.

There was a time when all the independent charter airlines suffered an image problem due to the fact that many went out of business with depressing regularity and those that did survive offered a lower standard of service. This may have been true 25 years ago, but one only has to look at the balance sheets to see that for many years Dan-Air always made a profit, and when it did not, there were external circumstances outside of it's control that played a large part in their losses.

A number of nick-names...

Over the years at least three nick-names -'Desperate' or 'Dangerous Dan' and 'Dan Dare' - and numerous derogatory quasi-advertising slogans have been applied to the company.

Interestingly, the nicknames appear to have their roots in a pair of cartoon characters from two boys' comics of the 1950s. Desperate Dan was a strongman cartoon strip in the '*Beano* '. The 'Desperate Dan' tag obviously relates to the lack of reliability, but 'Dangerous'? The airline's impressive record on safety and training for all its staff clearly speaks for itself.

As for the 'Dan Dare' tag, obviously this was derived from the Sci-Fi cartoon strip - a then futuristic, but so British 'stiff upper lip' Officer in Space Corps uniform - in the '*Eagle* ' comic, and it is clear that all were derived from the original characters title's similarity with the airlines name. But where is there a link between these characters and an independent airline? No matter which way it is viewed, there is not one. However, the tags stuck. Although referring to the airline as 'Dan Dare' was like waving a red rag to a bull to many of the staff, it was noticeable that when it came from passengers it was used (and received) almost as a term of endearment. Whether you upset an employee or not depended very much on the conversations context and your tone of voice...

Back in history Dan-Air may have slightly deserved the

...to 1991. The 1992 data is unaudited interim figures up to 30 June 1992.

1978	1979	1980	1981	1982	1983	1984	1985	1986	1987	1988	1989	1990	1991	1992
3407	3477	3258	3097	2957	2915	3108	3276	3657	3803	4264	4390	4325	2644	n/a
83	79	79	81	80	82	82	82	84	84	84	84	84	81	n/a
5299	4589	4373	4053	4775	4959	6219	6989	7775	8262	5809	6276	5893	4858	n/a
4010	3548	3510	3226	3599	3702	4599	5007	5309	5481	5809	6276	5893	4858	n/a
117,505	129,487	153,815	154,472	183,840	196,127	242,846	287,516	306,254	329,617	388,939	375,786	380,745	325,264	132.903
116,406	152,299	128,015	152,978	181,842	194,206	240,202	285,581	303,879	327,811	337,618	373,830	378,781	325,264	132.903
724	3,085	4,096	(95)	3,057	1,836	1,706	(108)	4,183	5,917	5,420	(1,856)	(25,928)	(21,669)	(39,050)
60	55	54	53	55	60	60	55	56	53	52	54	52	41	43

reputation, but to link it in print to the airlines latter years of operation was certainly libellous.

But there are others areas where the airline has carried the blame for events and occurrences that were certainly not its fault. Air Traffic Control delays cannot in any way be blamed on the airline, and yet with charter flights delayed for hours at airports around Europe, it was so easy for the phrase 'Time to spare - Fly Dan Dare' to come into being.

Tour operators regularly 'consolidate' - that is, bring a number of partially full flights together into one to make the most efficient use of an aircraft - at short notice. When passing such information on to holiday-makers, travel agents and tour representatives just tell them that 'the airline and departure time has been changed'. Whilst this is true, in their innocence the public perceive it as the airline's fault and it certainly must be easier for the travel company to let them think so, for it saves a lot of explanation!

As to who was seeing this 'poor brand image', it is difficult to say, for it certainly does not appear to be the passengers! During the research for this book the author received many letters and calls from recent travellers who provided totally unsolicited comments on today's operation...

'Peter Neighbour, Monaco.
"...I currently fly regularly from Nice to Gatwick, and at present, Dan-Air is ahead of the pack of Air France, BA, Air Canada, Air UK, British Midland etc., largely because of the consistently high standard of service, and a very good underlying sense of normality/humour or whatever.'

Nola & Peter Garrett, Bury, Lancs.
On our recent flight to Malaga my wife and I were reading the In Flight magazine with interest......we had not realised that the airline had been going since the early 1950s. To remain independent for such a long time in these times demands praise. It is a commercial achievement and makes mockery of such phrases as 'Dan Dare'. The jokes are wrong and without foundation'

And finally, one letter that deserves reproducing in its entirety:
'Anne Cullerton, Bradford.
Some five years ago I flew by Dan-Air to Alicante. Last week I returned by Dan-Air from two weeks in Skiathos, Greece.
In the interim I viewed on TV a programme regarding the new image that the Company was now projecting to drag it away from the Dan Dare label. I now feel strongly motivated to let you know from a passenger's point of view that your hard work has been well worthwhile.
The services and refreshments were excellent and your rivals only differ by the provision of a menu card.
Although not one of their number, your shareholders must be delighted - congratulations!'

So, it appears that many of the passengers were happy! One suspects that continued use of the phrases was in the media's own best interest; after all, they seem to like knocking anything that is a success and it is too easy to make use of a nick-name for a cheap laugh - it makes for attention grabbing headlines too!

A PERSONAL VIEW...

Year	DC3	York	Freighter	Heron	Dove	Ambi	DC4	DC7	Comet	1-11	262	Viscount	Herald	707	748	727	737	Airbus	146	Total
1953	1	-	-	-	-	-	-	-	-	-	-	-	-	-	-	-	-	-	-	1
1954	2	-	-	-	-	-	-	-	-	-	-	-	-	-	-	-	-	-	-	2
1955	2	1	-	-	-	-	-	-	-	-	-	-	-	-	-	-	-	-	-	3
1596	2	4	-	-	-	-	-	-	-	-	-	-	-	-	-	-	-	-	-	6
1957	2	4	1	1	-	-	-	-	-	-	-	-	-	-	-	-	-	-	-	8
1958	2	6	2	1	-	-	-	-	-	-	-	-	-	-	-	-	-	-	-	11
1959	2	4	3	-	-	3	-	-	-	-	-	-	-	-	-	-	-	-	-	12
1960	2	4	3	-	2	3	-	-	-	-	-	-	-	-	-	-	-	-	-	14
1961	3	4	3	-	2	3	-	-	-	-	-	-	-	-	-	-	-	-	-	15
1962	3	4	3	-	2	4	-	-	-	-	-	-	-	-	-	-	-	-	-	16
1963	3	2	3	1	2	6	-	-	-	-	-	-	-	-	-	-	-	-	-	17
1964	4	1	2	1	1	7	-	-	-	-	-	-	-	-	-	-	-	-	-	16
1965	4	-	2	1	1	7	2	-	-	-	-	-	-	-	-	-	-	-	-	17
1966	4	-	2	-	-	8	-	1	2	-	-	-	-	-	-	-	-	-	-	17
1967	3	-	2	-	-	6	-	1	4	-	-	-	-	-	-	-	-	-	-	16
1968	3	-	2	-	-	4	-	1	4	-	-	-	-	-	-	-	-	-	-	16
1969	1	-	1	-	-	3	-	1	11	3	-	-	-	-	-	-	-	-	-	20
1970	1	-	1	-	-	3	-	-	12	4	1	-	-	-	-	-	-	-	-	22
1971	-	-	-	-	-	1	-	-	14	5	1	-	-	1	2	-	-	-	-	24
1972	-	-	-	-	-	-	-	-	19	5	1	-	-	2	6	-	-	-	-	33
1973	-	-	-	-	-	-	-	-	22	5	-	-	-	2	7	3	-	-	-	39
1974	-	-	-	-	-	-	-	-	14	5	-	-	-	2	7	5	-	-	-	33
1975	-	-	-	-	-	-	-	-	19	13	-	3	-	2	9	5	-	-	-	51
1976	-	-	-	-	-	-	-	-	18	14	-	-	-	4	11	6	-	-	-	53
1977	-	-	-	-	-	-	-	-	17	14	-	-	-	6	17	8	-	-	-	62
1978	-	-	-	-	-	-	-	-	13	14	-	-	-	5	19	8	-	-	-	59
1979	-	-	-	-	-	-	-	-	7	15	-	2	-	1	21	8	-	-	-	54
1980	-	-	-	-	-	-	-	-	4	15	-	2	-	-	20	10	2	-	-	53
1981	-	-	-	-	-	-	-	-	-	17	-	1	-	-	21	11	2	-	-	52
1982	-	-	-	-	-	-	-	-	-	19	-	-	-	-	19	12	4	-	-	54
1983	-	-	-	-	-	-	-	-	-	20	-	-	-	-	18	11	8	-	2	59
1984	-	-	-	-	-	-	-	-	-	19	-	-	1	-	18	11	7	-	3	59
1985	-	-	-	-	-	-	-	-	-	20	-	-	-	-	14	9	8	-	4	55
1986	-	-	-	-	-	-	-	-	-	19	-	-	-	-	14	10	8	2	3	56
1987	-	-	-	-	-	-	-	-	-	18	-	-	-	-	12	10	8	1	5	54
1988	-	-	-	-	-	-	-	-	-	19	-	-	-	-	11	10	10	2	3	55
1989	-	-	-	-	-	-	-	-	-	17	-	-	-	-	10	12	11	2	4	56
1990	-	-	-	-	-	-	-	-	-	18	-	-	-	-	9	10	12	1	5	55
1991	-	-	-	-	-	-	-	-	-	15	-	-	-	-	6	7	16	-	5	49
1992	-	-	-	-	-	-	-	-	-	12	-	-	-	-	4	7	19	-	4	46
1992	-	-	-	-	-	-	-	-	-	10	-	-	-	-	-	6	18	-	4	38

Analysis of the Dan-Air Services Ltd fleet on an annual basis.
These figures do not take into account aircraft obtained as spares sources and machines used on joint services. They do include machines used by Dan-Air Engineering and aircraft on lease.

DAN-AIR SERVICES

With the benefit of hindsight...

There have been numerous press reports relating to the ongoing story of the Dan-Air/British Airways merger. As far as this book is concerned, these aspects are outside the subject: Dan-Air, in its independent form, ceased to exist with the hand-over at 2359 hrs on 8 November.

Since 23 October 1992 I have been asked many times one particular question - Why - what went wrong? Deep down, there was (and still is) much anger, for many people, including a large number of employees who had faith in their own survival, lost large amounts of money and, understandably perhaps, they were looking for a reason and some-one to blame. Some, I am sure, look towards this book to provide a simple answer.

I have turned that question over and over in my mind, talked to as many people as possible and looked at as many documents as I can, but I cannot find one simple answer to settle their minds. Instead, I think there was a myriad of reasons, circumstances and events that inexorably led up to the news that day.

Initially, there was a 'knee-jerk' reaction that cast British Airways in the role of 'villain'. I do not, nor have I ever thought that was the case. They had a number of sound commercial reasons for wanting to obtain the airline: they wanted its slots, its routes and its expertise. In concluding the deal, British Airways played a brilliant hand; their timing and the pressures exerted in concluding the deal could not have been bettered. No, the reasons go back further than the last few months.

Others have blamed David James, the second and last Chairman of Davies & Newman and Dan-Air. However, with the information and backround details provided by him during my extended interview on 18 January 1993, I am by no means sure that there are grounds for that belief, although I can see how people arrived at that conclusion. At the time the events were taking place, there was no alternative but to portray the Company's status in the best possible light, both internally and to a wider audience; anything else would have created a severe crisis of confidence and brought forth an uncontrollable collapse - exactly the thing to be avoided.

From the material provided at that interview, it is clear that Mr James and the remainder of the Board appeared to do all they could to, at best, preserve the airlines independence, or at worst, save the Company from recievership.

In hindsight, many people have thought that Dan should have arranged a 'marriage' with a medium sized tour operator to feed them seats for their charter flights. But, given the number of failures and amalgamations of tour operators in the late 1980s and early 1990s, an almost equal number of people have suggested that independence was the best thing, for otherwise Dan-Air could have

A PERSONAL VIEW...

found itself swallowed up or, remembering the spectacular collapse of Clarkson and its airline Court Line, be bled dry.

A case can be made for saying that the overall aircraft fleet was far too diversified, incurring higher costs than necessary on training and maintenance. But the range of seats that could be offered by the diverse nature of the aircraft interiors meant that flights could be tailored to suit the customer's needs. Not altogether a bad thing when fighting for custom!

Perhaps Dan-Air did embark on too-rapid an expansion into the scheduled service market, and from Gatwick, regarded by many as the wrong location for such a scheme. But realistically, there was no way they could ever move into the already chronically overcrowded Heathrow on the scale required to have an impact and attract the business world in droves. To generate the high levels of business traffic required from Gatwick would have required a fundamental change in the minds of the travelling public, who still see Gatwick as a 'holiday' airport. Despite this, John Olsen and his team did generate much business traffic to and from Gatwick. Perhaps the move to concentrate on Gatwick as a 'hub' was ill-advised. The Newcastle and Manchester scheduled service operations at their height were a good profit base for Dan and perhaps should have been built upon more. Nevertheless, it did make sense to 'go scheduled', for down that path lay possible year-round profits, not the 'peaks and troughs' of the charter side. It is far too easy to profer caution with the benefit of all-seeing hindsight.

Where did the fall start, and could it have been avoided?. With hindsight, I do not think so, for there were far too many external influences. Without doubt the decline started in the late 1980s, but was reaching a climax by 1990, and was brought about by a number of factors:

* The number of medium sized tour operators that established their own airlines, a market on which Dan-Air had traditionally relied and which created excess capacity for seats on the charter side of the operation. This excess was not helped by the recession that appeared at virtually the same time, for then the market was literally awash with vacant seats.

* The Gulf conflict had a large impact on the travelling plans of many from the business world; these were people who preferred to travel 'Class Elite' and thus generated high revenue per seat for the airline.

* Air Europe was in financial troubles and, in order to get out of its difficulties, quite legitimately hit out at Dan-Air, its major competitor by attacking it in the press and suddenly withdrawing all its work.

As a result, the airline started to experience financial troubles. Without doubt, the 1990 rescue package was justified. The airline was worth saving and there appeared the possibility that this could be done, on the one hand by reducing its costs with a streamlining of its operation - selling off the Engineering Division and aspects of

DAN-AIR SERVICES

its Airport Handling side for example - which greatly assisted in reducing its debts and, on the other, by firmly entering the scheduled service market to generate a smoother cash-flow for the Company. However, I suspect that the correct balance between sheduled and charter work was never fully established. 1991 and 1992 brought further problems that were, in many respects outside the airline's control:

* The Gulf War peaked in early 1991 and thus brought a severe cut-back in the travelling public's plans.

* The recession bit deeper and brought gloom to business and also meant that travelling plans for many were pared to the bone.

* The run-up to the 1992 election saw much speculation within the press of a Labour Government and therefore a potential loss of confidence in the business community. Again, this brought a threat of cut-backs in travel plans.

* The financial pressures on sterling during 1992 eventually lead up the a crisis and withdrawal of the pound from the Exchange Rate Mechanism caused great difficulties for any company that conducted business in foreign currency. It must be remembered that the international method of purchasing aviation fuel (or any oil) is in US dollars. The collapse of the pound against the dollar saw the value of the cash set aside by Dan-Air to pay its dollar debts greatly reduced.

David James has strong thoughts on the matter and, considering that his ideas are based on the knowledge of primary data, they are probably the most accurate and worth repeating here for consideration:

"If I was to look back now and say what Dan-Air (or Davies & Newman) should have done, it would be that I think they should have kept the charter business going; but essentially created or added a tour company of their own by the mid 1980s. As for scheduled services, they should have just cherry-picked the good routes - Paris Manchester etc, and kept the structure of those fat routes going. They should never have had a major dependence upon the scheduled work. What should have happened was to use their own charter capacity to build their own tour operation. They should have bought a tour company when they had cash in the bank - after all, this was a cash-rich organisation until 1989. There was so much that could have been done. Instead the balance got out of hand.

By the time I got hold of it, the airline was so wholly committed to being a scheduled network operator; we could not get out of the scheduled side and there was insufficient charter to compensate.

Then there was the problem of incompatible equipment. If I have one overriding regret, when I look back to the fund-raising of 1991, if I had got another £10 million at that time, (and I could have done, because we had gone for £42 million, but the City over-offered) but I was not able to accept the cash because we had declared in advance how much we were going for.

If I had been able to find a way of getting an extra £10 million of that in, and had the nerve to declare that I was going to use it to get rid of the 727 fleet then (which would have cost £9 million) I would have solved one of the worst problems that we had. We were actually flying those 727s just to cover the cost of flying them; they were net losers. We were having to fly routes just to provide services just to use the 727s - they were a dreadful burden.

Even so, I still believe that the business plan backing the capital raising in October 1991 was a feasible proposition at the time. Remember that this was a period where there was optimism within the civil aviation industry that the Gulf

A PERSONAL VIEW...

War had just been a glitch in the growth of passenger traffic and we would be back to normal in 1992.

Instead, the market started to dip again in the month before Christmas 1991 and never began to approach the industry's expectation for recovery and growth in the first half of 1992. We had obviously allowed some latitude in our cash projections to cope with market fluctuations, but certainly not at a level that would have compensated for the aggregate shortfall in revenue we suffered, particularly during the second and third quarters of 1991. Indeed, I very much doubt that we would have raised cash in October 1991 at the level which would have provided us with an adequate hedge against such a disastrous market out-turn.

Had we had enough financial fuel in the airline's tanks to fly though 1992, then I still believe that the fundamentals of our strategy were sound and that Dan-Air would have found its way back to recovery and prosperity from 1993 and onwards.

A half-decent market in the year following the re-financing was an essential pre-condition and the reality was far from this. In retrospect, this meant Dan-Air's tragedy was that we simply did not have the depth of reserves to bridge the transformation of the old Dan-Air into the new and revitalised airline that we so much wished to create for the future".

Many people have commented that perhaps the 'illustrative projection' from Coopers & Lybrand that backed up David James' campaign to raise finance a second time was over-optimistic. Again, with the benefit of hindsight it may appear that way, but at the time...?

As David James commented in 1993,"*Those projections represent the most trawled-through, worked-over set of statistics I have ever worked on in my life*".

In the end the numbers would not match, and a case could be made that it may well have been better to have let the airline die in October 1991. This would have saved the investors £53.75 million, but then maybe the banks would not have got their loans back!

My own opinion is that it was no-ones fault, just a whole set of mainly external circumstances that conspired against the airline's survival. Perhaps the banks should have followed the spirit of Dan-Air and had that bit more faith in the Company!

Now, having read this book and studied all the information available to the author, it is down to individual readers to make up their own minds...

And so we travel full circle...

It was that last ITN newcast, shown a little after 10.00 pm, that was later to provide the only relief to the whole sad day of 23 October 1992. Independent Television News' Picture Library had 'borrowed' what was possibly the earliest two black and white photographs (in the historical sense) from the authors collection - one of a DC-3 and the other showing one of Dan-Air's first 'hosties'. The pictures were only on the television screen for about five seconds each.

During the research for this book, literally hundreds of photographs had been obtained from many different sources, almost all from outside the company. The difficulty facing the author in using much of this material was that there were few recorded details as to its subject matter - it was going to be difficult to write captions to go with the photographs. When dealing with historical material, this sort of problem is quite common, and is reasonably easy to solve as far as the aircraft are concerned, for there are ways to identify the machine, its location and roughly the time period. But this is not the case for people - how does one put missing names to faces?.

One way was decided upon by the author - talk to as many people as possible within the Company and hope that someone remembers - but that one shown on ITN had defied all identification. It was obviously early in the Company's history, for the girl was standing on the rear steps in front of what appeared to be a Bristol Freighter. But who was she? No-one I spoke to could remember. After eight months I was beginning to be haunted by that smiling face and had more or less given up hope of ever finding out who she was.

Over the weekend I received a tremendous number of calls of support for the book to continue and a number of calls from the Press, which were fended off. Monday saw me still trying to find out exactly what was happening.

At 9.20 that morning, the phone rang. A female voice said she had got my number from ITN's Picture Library and asked if I was the Mr Simons that had provided the pictures of Dan-Air to ITN the previous Friday evening. My reaction was to think that this was the press on the scrounge for more photo's. I was about to refer the speaker to Dan-Air's PR Department and then hang up, when the caller said she was very interested in the photograph of the hostie. That made me pause, for the word 'hostie' is airlinespeak, not the sort of thing that the press would know. Her next sentence left me staggering. *"I'm Andrea Charlwood, and that's a picture of my mum Maurene!"*

And so, right at the very sad end of an era that had provided so much to so many, the story had literally gone full circle. A report that recorded the demise of a great airline, also provided the spark that linked the very beginning, for Maurene was one of the first 'Dan-Air Girls'.

The 'Spirit of Dan-Air' had struck again - still defiant, still friendly, still providing a personal service and still coming up with the goods in the face of adversity. I should have followed Rosemary Cooper's advice when she told me *"You've got to have faith"*, for that was really the 'Spirit of Dan-Air'.

Appendix One
The Directors of Dan-Air Services Ltd

Over the 40 years of Dan-Air Services Ltd history, there has been remarkably few Directors. The list is in alphabetical order.

Mr. D. L. Bernstein.
Mr. L C. Crockford.
Mr. W. J. Crosby.
Mr. E. M. Davies.
Mr. J. W. Davies.
Mr. J. E. Denman.
Mr. E. T Evans.
Mr. J. Fitzgerald.
Mr. D. F. Frost.
Mr. D. P Herbert.
Mr. F. Horridge.
Mr. E. G. Hutchinson.
Mr. D. N. James.
Mr. W. Jones.
Mr. K. V. Kellaway.
Mr. A. H. Larkman.
Mr. J. E. C. Mayes.
Mr. L. G. Moore.
Mr. F. E. F. Newman.
Mr. M. R. F. Newman.
Mr. H. B. Olsen.
Mr. B. M. O'Regan.
Sir Ian M. Pedder.
Mr. J. G. Penwarden.
Mr. R. A. Pigeon.
Mr. S. N. Poulton.
Mr. C. E. Powell.
Mr. T. J. Redburn.
Mr. G. C. Ruffle.
Mr. P. H. Ryan.
Mr. V. K. M Sheppard.
Mr. D. J. Siddaway.
Mr. R. J. Smith.
Mr. A. J. A. Snudden.
Mr. F. G. Tapling.
Mr. J. W. Varrier.
Mr. E. W. Wallis.
Mr. C. R. Whyte.
Mr. B. V. S. Williams.

DAN-AIR SERVICES

List of Company Secretaries

Mr A. F. Austin.
Mr. D. R Silburn.
Mr. J. A Brown.
Mr. J. E. M. Gardner.
Mr. A Garretts.
Mr. J. A. Brown.
Mr. B. M. O'Regan.
Mr. D. P. Herbert.
Mr. D. J. Quinn.

GATWICK HANDLING LTD.
The list of Directors is in alphabetical order.

James Abbott.
Malcolm Bromham.
Sir Ivor Broom.
Chris Brown.
Sir Robert Hardingham.
John P Harvey.
Roger D Hauge.
Frank Horridge.
Graham Hutchinson.
Allan Johnson.
Angus J Kinnear
Egon Koch.
Cliff Nunn.
Michael Newman.
Peter J O'Boyle.
Martin O'Regan.
Allan K Pray.
Steven Rothmeier.
Peter Ryan.
John Trayton Seear.
Alan Snudden.
David J Quinn.
Harold Whincup.
Charles R Whyte.

DAN-AIR SERVICES

Appendix Two
The aircraft of Dan-Air Services Ltd

Throughout the forty years of operations celebrated by the publication of this book, Dan-Air Services used and flew a considerable number of different aircraft. In order to make this proliferation of types, sub-types and variants easier to understand, (and put their use into some form of perspective), from the files of Peter Carter, Commercial Manager, Aircraft comes a complete fleet list of all aircraft owned or operated by Dan-Air Services

To allow the reader easy access to the data, the following details are listed in aircraft type sections, within these sections aircraft are listed by registration, construction number, mark, date into service, acquired from, disposal date/who to.

Douglas C47B Dakota

G-ALXK	32828	4	Delivered 2/64 from Capital Finance Co. Last service flown 11/10/66. Scrapped 11/69.
G-AMPP	26717	4	Delivered 3/61 from Scottish Airlines Ltd. Last service flown 9/70. Preserved Lasham as G-AMSU.
G-AMSS	32840	4	Obtained 2/54 from William Dempster Ltd. Last service flown 11/68. Sold to Air Taxi Ltd, Tehran.
G-AMSU	33548	4	Obtained 5/53 from Meredith Air Transport. Last service flown 24/3/68. Scrapped 1972.

Avro 685 York

G-AMUT	-	-	Ex-RAF, serial MW185. Obtained from Maritime Central Airways of Canada March 1958. Damaged beyond repair Luqa, Malta 20/5/58.
G-AMUV	-	-	Ex-RAF, serial MW226/XD623/XD669. Obtained from Air Charter, Stansted 2/56. Crashed in forced landing due to engine fire Gurgaon, India 25/5/58.
G-ANTI	-	-	Ex-RAF, serial MW143. Obtained from Ministry of Defence 7/54 & registered 18/1/55. Last service flown 19/1/62 & scrapped Lasham 5/63.
G-ANTJ	-	-	Ex-RAF, serial MW149. Obtained from Ministry of Defence 7/54 & registered 23/12/58. Last service flown 26/11/62 & scrapped Lasham 12/62

DC-3 G-AMPP, seen at Liverpool on 5 July 1966. Note the cabin staff and their blue uniforms. (M Forsyth)

DAN-AIR SERVICES

York G-ANXN operating a BEA freight flight from Heathrow.
(M Forsyth Collection)

G-ANTK - - Ex-RAF, serial MW232. Obtained from Ministry of Defence 7/54 & registered 30/10/56. Last service flown 30/4/64. Withdrawn from use Lasham 10/64. Preserved. To Imperial War Museum, Duxford.

G-ANXN - - Ex-RAF, serial MW258. Obtained from Air Charter, Stansted 9/56. Last service flown 26/2/63 & scrapped Lasham 6/63.

Bristol 170 Freighter

G-AINL 12827 Mk.31. Obtained from Aviation Traders Ltd (Bristol Aeroplane Co.) 6/57. Last service flown 22/2/70. Sold to Lambair of Canada as CF-YDO 2/70.

G-AMLL 13074 Mk.31. Obtained from Jersey Airlines Ltd 7/59. Last service flown S.2/63. Sold to Handley Page Ltd in exchange for two Ambassadors. Sold to Air Ferry Ltd 1964.

G-APLH 13250 Mk.31. Delivered from Bristol Aeroplane Co. 31/3/58. Last service flown 10/68. Sold to Lambair of Canada as CF-YDP 10/68.

de Havilland 104 Dove

G-AIWF 04023 Mk.1b. Delivered 4/1/60 from College of Aeronautics. Last service flown 19/2/65. Sold to Keegan Aviation Ltd.

G-ALVF 04168 Mk.1b. Delivered 5/2/60 from Hants & Sussex Aviation. Last service flown 25/10/63. Sold 10/63 to Cameroon Airways.

de Havilland 114 Heron

G-ANCI 14043 Mk.1b. Delivered on lease from Overseas Air Transport 6/57. Last service flown when returned 1/58.

G-AOZM 14002 Mk.1b. Obtained from Itavia, ex-I-AOZM 29/4/63 via Keegan Aviation. Named *'Plymouth Flyer'* Last service flown 27/9/65. Sold to Air Services Ltd, (also listed as Avitour) Isreal as 4X-ARK 9/65.

Airspeed A.S.57 Ambassador

G-ALFR 5210 Mk.2. Obtained from Napier Engines 3/61. Not placed into service until 1964. Last service flown 11/4/66. Withdrawn from use & scrapped Lasham 3/67.

G-ALZN 5212 Mk.2. Obtained from Overseas Aviation (CI) Ltd 4/62. Last service flown 25/9/67. Withdrawn from use & scrapped 5/68.

G-ALZO 5226 Mk.2. Obtained from Handley Page Ltd 20/2/63. Last service flown 26/9/71. Preserved Lasham. To Imperial War Museum, Duxford.

G-ALZR 5214 Mk.2. Fitted with freight door. Obtained from BKS Air Services 11/69. Not flown, scrapped 1972.

G-ALZX 5220 Mk.2. Obtained 11/59 from Butler Air Transport ex-VH-BUI. Damaged beyond repair on landing at Beauvais 14/4/66. Scrapped Beauvais.

- THE AIRCRAFT

A beautiful air-to-air photograph of Ambassador G-AMAE above the clouds.
(DAS/GMS Collection)

G-ALZY	5221	Mk.2.	Obtained from Handley Page Ltd 20/2/63. Last service flown 10/11/67. Withdrawn from used & scrapped Lasham 1967.
G-AMAA	5223	Mk.2.	To Dan-Air at Lasham 5/66 from Shell Aviation Ltd as spares. Not flown, & scrapped 2/67.
G-AMAE	5227	Mk.2.	Obtained 11/59 from Butler Air Transport ex-VH-BUK. Last service flown 30/9/70. Scrapped 1972.
G-AMAG	5229	Mk.2.	Obtained from Shell Aviation Ltd 5/66. Damaged beyond repair when landing at Manston 30/9/68. Scrapped.
G-AMAH	5230	Mk.2.	Obtained 11/59 from Butler Air Transport ex-VH-BUJ. Last service flown 30/10/70. Scrapped 1972.

Miles (Handley Page) H.P.R.1 Marathon

G-AILH	6430		Obtained 9/58 from Royal Air Force, but never entered service. Scrapped 5/59.

Douglas C54A (DC-4) Skymaster

G-APID	10408	Mk.a	Leased from Trans World Leasing 2/65. Returned 10/65.
G-ARXJ	18370	Mk.b	Leased from Trans World Leasing 3/65. Returned 10/65.

Douglas DC-7 CF

G-ATAB	45361	CF	Obtained from Trans World Leasing 3/66. Last service flown 25/6/69. Scrapped 1972.

DC-7F G-ATAB on the scrap-heap at Lasham after retirement. This aircraft was nicknamed by the Dan-Air crews 'The Torrey Canyon' (for it was always leaking oil) after the oil tanker that ran aground on rocks off the British coast.
(M Forsyth Collection)

227

DAN-AIR SERVICES

Comet G-AROV climbs away from Gatwick for another trip to the sun. (DAS/GMS Collection)

De Havilland DH106 Comet

G-APDA	06401	Mk.4.	Obtained from Malaysia-Singapore Airlines 11/69, ex-9V-BAS. To Lasham for spares - not flown. Scrapped 9/72.
G-APDB	06403	Mk.4.	Obtained from Malaysia-Singapore Airlines 16/10/69, ex-9V-AOB. Last service flown 28/11/73 & presented to East Anglian Aviation Society. Later to Duxford Aviation Society. Preserved on static display.
G-APDC	06404	Mk.4.	Obtained from Malaysia-Singapore Airlines 29/8/69, ex-9V-BAT. Last service flown 10/4/73. Scrapped 4/75.
G-APDD	06405	Mk.4.	Obtained from Malayan Airways 10/69, ex-9M-AOD. To East African Airways on lease during 1970 as 5Y-AMT. Last service flown 24/8/72. Scrapped 3/73.
G-APDE	06406	Mk.4.	Obtained from Malaysia-Singapore Airlines 11/69 ex-9V-BAU. To East African Airways on lease during 1970 as 5Y-ALF. Returned 1/71. Last service flown 2/4/73. Scrapped 4/73
G-APDG	06427	Mk.4.	Obtained following lease from BOAC to Middle East Airways 6/3/70. Obtained from Kuwait Finance Corp. Last service flown 2/5/73. Scrapped 6/74.
G-APDJ	06429	Mk.4.	Obtained on lease from BOAC 10/67. Sold to Dan-Air 6/72. Last service flown 28/11/72. Scrapped 6/74.
G-APDK	06412	Mk.4.	Obtained from BOAC 19/5/66. Leased to East African Airways 2/70 as 5Y-ALD. Returned 3/70. Last service flown 7/5/73 & presented to Lasham Air Scouts. Scrapped 9/80.
G-APDL	06413	Mk.4.	Obtained from BOAC 7/1/69. Written off 7/10/70 during wheels up landing at Woolsington.
G-APDM	06414	Mk.4.	Obtained from Middle East Airlines following lease to Malayan Airlines ex-9V-BBJ 3/69. Last service flown 8/10/73. WFU 5/74 & used for cabin crew training at Gatwick.
G-APDN	06415	Mk.4.	Obtained from BOAC 10/67. Crashed in mountains near Barcelona 4/7/70.
G-APDO	06416	Mk.4.	Obtained from BOAC 26/5/66. Last service flown 22/6/73. Scrapped 6/74.
G-APDP	06417	Mk.4.	Obtained from BOAC 3/69 after lease to Malayan Airlines 1/69 ex-9V-BBH. Last service flown 22/3/73. To Ministry of Defence (RAE Farnborough) for Airborne Early Warning Nimrod 6/73.
G-APMB	06422	Mk.4b	Obtained from Channel Airways Ltd 9/4/72. Last service flown 28/12/78. Sold to Gatwick Handling.
G-APMC	06423	Mk.4b	Obtained 10/73 from BEA Airtours. 2/11/73 Never entered service. Scrapped 4/74.

- THE AIRCRAFT

Reg	C/N	Mark	Notes
G-APMD	06435	Mk.4b	Obtained 12/9/72 from BEA Airtours. Withdrawn from use 1/1975. Scrapped 1978.
G-APME	06436	Mk.4b	Obtained 24/2//72 from BEA. Last service flown 2/5/78. Scrapped 6/79.
G-APMF	06426	Mk.4b	Obtained 30 1/73 from BEA Airtours. Last service flown 11/74. Scrapped 1976.
G-APMG	06442	Mk.4b	Obtained from BEA Airtours 19/1/73. Last service flown 21/11/77. Scrapped 15/4/78.
G-APYC	06437	Mk.4b	Obtained from Channel Airways Ltd 4/72. Force landed at Manston on 19/5/73 with nosewheel failure. 7 Crew & 110 passengers aboard. Repaired. Last service flown 4/12/78. Sold to Ministry of Defence, Kemble.
G-APYD	06438	Mk.4b	Obtained from Channel Airways Ltd 4/72. Last commercial flight 23/10/79. Presented to Science Museum, Wroughton 31/10/79 for preservation.
G-APZM	06440	Mk.4b	Obtained from Channel Airways Ltd 17/4/72. Last service flown 14/11/78. Scrapped 9/80.
G-ARCP	06431	Mk.4b	Obtained 10/73. Re-registered G-BBUV 12/73.
G-ARDI	06447	Mk.4b	Obtained from Channel Airways Ltd 4/72. Scrapped Stansted 6/72. Never entered service.
G-ARGM	06453	Mk.4b	Obtained 10/73 from BEA Airtours 1/11/73. Never entered service. Scrapped 9/74.
G-ARJK	06452	Mk.4b	Obtained from BEA Airtours Ltd 5/10/73. Last service flown 1/11/76. Scrapped 10/77.
G-ARJL	06455	Mk.4b	Obtained from BEA Airtours 9/11/73. Never entered service. Scrapped 9/74.
G-ARJN	06459	Mk.4b	Obtained from BEA Airtours Ltd 15/2/73. Last service flown 24/12/77. Scrapped 11/78.
G-AROV	06460	Mk.4c	Obtained from Aerolineas Argentinas 20/10/71 ex-LV-AIB. Last service flown 31/10/77. Scrapped 11/78.
G-ASDZ	06457	Mk.4c	Obtained from Sudan Airways ex-ST-AAW 2/6/75. Never entered service. Scrapped 1975
G-AYVS	06474	Mk.4c	Obtained 24/3/71 following lease to Middle East Airways from Kuwait Airways 9/75 ex-9K-ACE. Last service flown 6/1/77. Scrapped 4/78.
G-AYWX	06475	Mk.4c	Obtained 30/3/71 following lease to Middle East Airways from Kuwait Airways 3/71 ex-9K-ACA. Last service flown 2/5/78. Scrapped 10/79.
G-AZIY	06434	Mk.4	Obtained from Aerolineas Argentinas 8/11/71 ex-LV-AHU. Last service flown 25/11/73. Scrapped 3/77.
G-AZLW	06432	Mk.4	Obtained from Aerolineas Argentinas 11/71 ex-LV-AHS. Never entered service. Scrapped 1/73.
G-BBUV	06451	Mk.4b	Obtained from BEA Airtours 22/10/73. Last service flown 23/10/78. Withdrawn from use Lasham. Scrapped 10/79.
G-BDIF	06463	Mk.4c	Obtained by Dan-Air from Sudan-Airways ST-AAX 2/6/75. Last service flown 5/11/79. Scrapped 10/80.

Comets all!
Seen at Gatwick is this line of four ('YC closest) of De Havilland's famous airliner. Just visible at the left end of the line is the nose of a company BAC 1-11.
(DAS/GMS Collection)

DAN-AIR SERVICES

G-APME, a Comet 4B somewhere in the Mediterannian in October 1974 (DAS/GMS Collection)

G-BDIT	06467	Mk.C4	Obtained 29/8/75, ex-216 Squadron, RAF, serial XR395. Last service flown 13/10/80. Sold to Warbirds of Great Britain Ltd, Blackbushe 8/6/81.
G-BDIU	06468	Mk.C4	Obtained 29/8/75, ex-216 Squadron, RAF, serial XR396. Last service flown 6/10/80. Sold to British Aerospace, Bitteswell 9/7/81.
G-BDIV	06469	Mk.C4	Obtained 29/8/75, ex-216 Squadron, RAF, serial XR397. Last service flown 12/11/79.
G-BDIW	06470	Mk.C4	Obtained 29/8/75, ex-216 Squadron, RAF, serial XR398. Last service flown 9/11/80. Sold to Air Classic Collection, Dusseldorf 7/2/81.
G-BDIX	06471	Mk.C4	Obtained 29/8/75, ex-216 Squadron, RAF, serial XR399. Last service flown 17/10/80. Sold to Museum of Flight, East Fortune, Scotland
G-BEEX	06458	Mk.4b	Obtained 14/10/76 ex-Egyptair SU-ALM. Never entered service. Scrapped 1977.
G-BEEY	06462	Mk.4c	Obtained 10/10/76 ex-Egyptair SU-ALV. Never entered service. Reduced to spares & scrapped Lasham 1977.
G-BEEZ	06466	Mk.4c	Obtained 16/10/76 ex-Egyptair SU-ANC. Never entered service. Scrapped 1977.
SU-ALL	06454	Mk.4c	Obtained from Egyptair 10/76. Scrapped Cairo 10/76.
5H-AAF	06433	Mk.4	Obtained from East African Airways 1/71. Flown from Niarobi to Lasham during 1/71. Never entered service. Dismantled for spares 1973.
5X-AAO	06431	Mk.4	Obtained from East African Airways 11/70. Flown from Niarobi to Lasham. Never entered service. Dismantled for spares 1971/2.
5Y-AAA	06472	Mk.4c	Obtained from East African Airways 2/71. Flown from Niarobi to Lasham. Never entered service. Dismantled for spares 1971.
OD-ADT	06450	Mk.4c	Obtained from Middle East Airlines 4/10/73. Scrapped 6/74. Never entered service.
LV-AHN	06408	Mk.4	Obtained by Dan-Air from Aerolineas Argentinas 12/71. Flown to Lasham. Never entered service. Dismantled for spares 1971.
LV-AHS	06432	Mk.4	Obtained from Aerolineas Argentinas. To Dan-Air 1/72. Flown to Lasham. Never entered service. Dismantled for spares 1971.

- THE AIRCRAFT

On the ramp at Gatwick awaiting the next load of passengers is Vickers Viscount G-BCZR (DAS/GMS Collection)

Handley Page Jetstream
G-AXEK 203 137 Demonstration aircraft. Evaluated by Dan-Air.

Handley Page Herald
G-BAVX 184 214 Obtained on lease from British Air Ferries 8/1/84. Last service flown 28/2/84. Returned from lease 1/3/84.

Vickers Viscount
G-ARBY 010 708 Obtained on lease from Alidair 10/75. Returned 12/75.
G-ARIR 036 708 Obtained on lease from Alidair 12/75. Returned 12/75.
G-BBDK 291 808C Obtained on lease from Air Bridge Carriers 3/75. Returned 10/75.
G-BGLC 436 810/839 Obtained 30/1/79 from Field Aircraft Services. Last service flown 26/10/80. To Air Zimbabwe 30/10/80.
G-BCZR 446 810/839 Obtained 19/3/79 from Swaziland. Last service flown 27/4/81. To Air Zimbabwe 1/5/81.

Nord 262
F-WNDD 40 A Manufacturers Aircraft. Flown in Dan-Air colours for evaluation.
G-AYFR 29 A Obtained from Air Ceylon, ex 4R-ACL 6/70. Last service flown 22/2/72. Sold to Rousseau Aviation as F-BTDQ 2/72.

Boeing 707
G-AYSL 17599 321 Obtained from Pan American Airways, ex-N721PA 6/3/71. Last service flown 16/10/79. Sold to Del Air Inc, Miami 26/10/79.
G-AZTG 17600 321 Obtained from Pan American Airways, ex-N722PA 5/5/72. Last service flown 14/11/78. Scrapped Lasham 5/82.
G-BEBP 18579 321C Obtained 30/6/76 ex Pan Am. Last service flown 14/5/77. Crashed Lusaka 14/5/77.
G-BEAF 18591 321C Obtained 21/5/76 ex Pan Am. Last service flown 5/7/78. To Compania Interamerica Export/Import Panama for TAR Argentina 7/7/78.
G-BEVN 19721 321C Obtained 7/7/77 ex Pan Am. Last service flown 30/9/78. Returned to Atasco 30/9/78.
G-BEZT 19765 321C Obtained 2/11/77 ex Pan Am. Last service flown 25/6/78. To Zimex, Switzerland 10/7/78.

DAN-AIR SERVICES

*G-BEVN, a Boeing 707-321C at Gatwick.
(DAS/GMS Collection)*

Hawker Siddeley (Avro) 748

G-ARAY	1535	200	Obtained 3/5/71 ex-OY-DFV from Maersk Air. Last service 17/10/89. Broken up Lasham
G-ARMW	1537	101	Obtained 7/4/72 from Skyways International. Leased to Air BVI as VP-LVO. Returned. Last service flown 7/9/90. Sold to Clewar Aviation.
G-ARMX	1538	101	Obtained 7/4/72 from Skyways International. Lease to Air BVI as VP-LVN. Returned. Sold to Air BVI 20/11/88.
G-ARRW	1549	106	Obtained 7/4/72 from Skyways International. Last service flown 20/3/86 Sold to Ryan Air 26/3/86.
G-ASPL	1560	108	Obtained 7/4/72 from Skyways International. Last service flown 26/6/86 when crashed Nailstone.
G-ATMI	1592	225	Obtained 13/5/75 ex VP-LIU from Court Line Receiver. Last service flown 3/5/91. Sold to Aberdeen Airways
G-ATMJ	1593	225	Obtained on lease from Civil Aviation Authority 5/75. Returned later that year. Also quoted as ex VP-LAJ, lease to British Airways. Purchased from CAA 21/7/78. Last service flown 17/10/91. In store Manchester To Janes Aviation, Blackpool.
G-AXVG	1589	226	Obtained 4/10/71 from Skyways International. Leased to Phillipines Airlines as RP-C1031 1985. Restored to Dan-Air. Last service 16/10/84. To Springer Airlines, Canada 20/6/89
G-AYYG	1697	225	Obtained on lease 17/6/78. Returned to Air New Zealand 8/10/78. Obtained on lease 4/79. Returned to Air New Zealand 17/9/79. Obtained on lease 9/4/80. Returned to Air New Zealand 22/10/81.

*G-ASPL, a 748 in the joint Dan-Air Skyways colours.
(DAS/GMS Collection)*

232

- THE AIRCRAFT

A beautiful air-to-air photograph of Avro 748 G-BEBA over the sea. (DAS/GMS Collection)

Reg	C/N	Series	Notes
G-AZSU	1612	232	Obtained 7/4/72 from BOAC Associated Companies. Leased as A2-ABB. Returned to Dan-Air. Last service flown 1/6/90. Sold to Aberdeen Airways
G-BEBA	1613	233	Obtained 17/6/76 from Air Pacific. Last service flown 29/10/84. Leased to to Phillipines Airlines as RP-C10321985. Sold to Springer Airlines 5/6/89.
G-BHCJ	1663	209	Obtained 12/10/79. Last service flown 22/10/84. To Phillipines Airlines as RP-C1033 1985. Sold to Springer Airlines 10/6/89
G-BEJD	1653	105	Obtained 17/12/76 ex LV-HHE of YPF Argentina. Cargo aircraft.
G-BEJE	1556	105	Obtained 4/2/77 ex LV-IDV of YPF Argentina. Cargo aircraft.
G-BEKC	1541	105	Obtained 10/2/77 ex LV-HHC of Aerolineas Argentina. Last service flown 21/12/87. Sold to Avocet.
G-BEKD	1544	105	Obtained 10/6/77 ex--LV-HHF of Aerolineas Argentina. Last service flown 8/83. Sold to Air Condal 9/83.
G-BEKE	1545	105	Obtained 24/3/77 ex LV-HHG of YPF Argentina. To Janes Aviation, Blackpool.
G-BEKF	1542	105	Obtained 24/3/77 ex-LV-HHD of YPF Argentina Scrapped Sumburgh 31/7/79 following runway accident.
G-BEKG	1557	105	Obtained 12/7/77 ex-LV-IEE Aerolineas Argentina. Last service flown 12/81. To Cayman Airways Ltd 23/2/82.

A dramatic photograph of 748 Kilo Echo as it comes into land in the Shetlands. (:DAS/GMS Collection)

233

DAN-AIR SERVICES

G-BFLL	1658	245	Obtained 10/2/78 from Avianca ex- HK-1409. Last service flown 30/4/90. Sold to Aberdeen Airways.
G-BIUV	1701	266	Obtained 27/5/81 ex 5W-FAN from Polynesian Airlines Converted to carry large freight door. To Janes Aviation, Blackpool.

Cessna 150

G-ATPM	0062	F	Obtained 8/74 from Safeflight Ltd. Crashed Whitchurch, Bristol 17/12/90.

Piper PA23 'Apache' 160

G-ATFZ	23-1314	-	Obtained from BAC Bavaria Flug Geselschaft. Crashed Godalming 1/9/66.

Cessna 310

G-AZVY	0040	Q	Previously SE-KFV. Operated 1982 in joint Dan-Air/Centreline colours.

Cameron Hot Air Balloon

G-BDWW	226	0-77	Obtained from Cameron Balloons. Crashed, Sussex 21/7/84

De Havilland Canada Twin Otter

G-BELS	300	300	Operated 1982 in colours of Metropolitan Airways on 'Link City' network.
G-BHFD	434	310	Operated 1982 in colours of Metropolitan Airways on 'Link City' network.

Embraer Bandeirante

G-CTLN	261	P1	Operated 1982 in joint Dan-Air/Centreline colours.
OY-ASL	165	P1	Operated 1982 in joint Dan-Air/Centreline colours.

BAC 1-11

G-ATTP	039	207AJ	Obtained from Zambian Airways, ex-9J-RCH 31/3/75. Last service flown 27/10/91. Sold to International Leasing & Finance Corp. (Ladeco Chile).
G-ATVH	040	207AJ	Obtained from Zambian Airways, ex-9J-RCI 3/75. Named "City of Newcastle". Last service flown 11/11/91. Sold to International Leasing & Finance Corp. (Ladeco Chile).
G-ATPJ	033	301AG	Obtained from Kuwait Finance Corp) 6/3/70. Last service flown 27/10/90. Sold to International Leasing & Finance Corp.

BAC 1-11-500 G-BEKA was rushed into service with just the addition of Dan-Air titles and Compass on the tail over the basic scheme of Arkia, its previous operators. As the Chairman liked the colours the new (and final one) was derived from this, with the red/blue bands reversed and lifted up the fuselage somewhat....
(M Forsyth Collection)

234

- THE AIRCRAFT

... as can be seen in this picture of 1-11 G-ATVH on approach to land. (DAS/GMS Collection)

Reg	MSN	Fleet	Details
G-ATPK	034	301AG	Obtained from on 15/3/85 from Chemco Bank. Last service flown 2/4/91. Sold to Okada, Nigeria.
G-ATPL	035	301AG	Obtained from Kuwait Finance Corp (also listed as British Eagle International Airways) 20/9/69. Last service flown 15/11/90. Sold to International Leasing & Finance Corp.
G-AVOE	129	416EK	Obtained on lease from British Aerospace 12/5/83. Last service flown 26/9/83. Returned from Lease 26/9/83.
G-AVOF	131	416EK	Obtained on lease from British Aerospace 21/5/84. Last service flown 15/10/84. Returned from Lease 3/85.
G-AXCK	090	401AK	Obtained from American Airlines Inc , ex-N5044 9/3/69. Last service flown 3/1/83. Sold to Westinghouse, USA 2/83.
G-AXCP	087	401AK	Obtained from American Airlines Inc, ex-N5041 9/3/69. Last service flown 31/10/86. Sold to British Aeropace 11/86
G-AXYD	210	509EW	Obtained from British Caledonian Airways 27/4/76. To Bournemouth 6/11/92 for disposal.
G-AZED	127	414EG	Obtained from British Aircraft Corp, ex-Bavaria Fluggesellschaft, D-ANOY 12/71. Sold to BAe 11/86.Obtained on lease from Florida Express 10/4/87. Returned off lease 4/11/87.
G-AWWX	184	509EW	Obtained from British Caledonian Airways 3/10/75. To Bournemouth 3/11/92 for disposal.
G-AZPZ	229	515FB	Obtained from Hapag Lloyd Fluggelschaft/British Caledonian Airways 1/5/81. Returned to British Caledonian Airways 28/2/82.
G-BCWA	205	518FG	Obtained from Court Line, (ex-G-AXMK '*Halcyon Star*') 24/1/75. To Bournemouth 6/11/92 for disposal.
G-BCXR	198	517FE	Obtained from British Aerospace ex G-BCCV 23/3/83. To Bournemouth 3/11/92 for disposal.
G-BDAE	203	518FG	Obtained from Court Line, (ex-G-AXMI '*Halcyon Day*') 28/3/75. Named '*The Highlander*' To Bournemouth 6/11/92 for disposal.
G-BDAS	202	518FG	Obtained from Court Line, (ex-G-AXMH '*Halcyon Sun*') 17/3/75. To Southend 5/11/92 for disposal.
G-BDAT	232	518FG	Obtained from Court Line, (ex-G-AYOR '*Halcyon Dawn*') 31/1/75. To Southend 6/11/92 for disposal.
G-BEKA	230	520FN	Obtained from Arkia, Isreal ex 4X-BAR 27/9/79. To Bournemouth 6/11/92 for disposal.
G-BJMV	244	531FS	Obtained on lease from Lineas Aereas Costarricenses

235

DAN-AIR SERVICES

			Costa Rica (LACSA) ex TI-LRJ 6/11/81. Returned off lease 28/10/82. Leased 28/10/82. To Bournemouth 2/11/92 for disposal.
G-BJYL	208	515FB	Obtained on lease from Lineas Aereas Costarricenses Costa Rica (LACSA) 12/3/82. To Bournemouth 2/11/92 for disposal.
G-BJYM	242	531FS	Obtained on lease from Lineas Aereas Costarricenses Costa Rica (LACSA) ex TI-LRI 11/5/82. To Bournemouth 2/11/92 for disposal.
G-BPNX	110	304AX	Obtained from Havlet Leasing 5/4/88 ex G-YMRU. Last service flown 29/11/88 when returned off lease.
G-BYSN	186	509EW	Ex-G-AWWZ. Obtained on lease from Atlantic International Holdings Inc 13/4/90. Last service flown 28/10/90. Returned off lease.
G-TARO	272	525RC	Obtained 6/84, ex YR-BCO on lease from Air Tarom, Romania. Last service flown 24/12/85. Returned off lease 28/12/85.
YR-BRD	404	561RC	Obtained on lease 17/9/88 from Tarom. Last service flown 10/10/88 when returned off lease. Leased again from 25/3/89 to 30/10/89.

A300 Airbus

G-BMNA	169	B4-203	Obtained on lease from Hapag Lloyd ex D-AHLJ 29/4/86. Last service flown 15/12/86. Returned off lease 18/12/86.
G-BMNB	009	B4-203	Obtained from Hapag-Lloyd, ex-D-AMAP 17/12/86. Last service flown 7/12/89. To Air Inter via GPA.
G-BMNC	012	B4-103	ex D-AMAX of Hapag Lloyd. Leased from Guiness Peat Aviation 23/8/88. Last service 15/10/90. Sold to Air Inter via GPA.

Boeing 727

G-BAEF	18879	46	Obtained from Japanese Air Lines, ex-JA8312 3/73. Leased to Royal Nepal Airlines as 9N-ABV 1985. Returned by 6/87. Last service flown 31/10/90. Parked in Mojave Desert 1/3/92.
G-BAFZ	18877	46	Obtained from Japanese Air Lines, ex-JA8310 3/73. Delivered 6/4/73. Leased to Avianca as HK-3270X then restored to company after lease 6/86. Last service flown 21/10/87. Sold to Aeron Corp 6/3/86.
G-BAJW	18878	46	Obtained from Japanese Air Lines, ex-JA8311 3/73. Sold to Amerijet USA 410/89. Last service flown 24/9/89.

*Airbus G-BMNB parked on the ramp 'somewhere in Spain' awaiting loading.
(DAS/GMS Collection)*

- THE AIRCRAFT

G-BCDA	19281	46	Obtained from Japanese Air Lines, ex-JA8320 4/74. Delivered 22/4/74. Last service flown 1/11/90. Sold to Aeron Aviation.
G-BDAN	19279	46	Obtained from TOA Domestic Airlines, ex-JA8318 8/74. Crashed & destroyed Tenerife 25/4/80.
G-BEGZ	19620	193	Obtained 20/10/76 ex Burma Airways Last service flown 7/4/81. To Aravco 10/4/81.
G-BFGM	19249	095	Obtained 22/12/77 ex Delta Air Lines. Last service flown 8/12/82. To Aeron Aviation Coporation 16/12/82.
G-BFGN	19251	095	Obtained 30/11/77 ex Delta Air Lines. Last service flown 15/10/81. To Airesearch USA 20/10/81.
G-BHNE	21676	2J4ADV	Leased from Sterling Airways, ex-OY-SBC. Delivered 20/3/80. Last service flown 12/10/91. Sold to Northern lakes Equity.
G-BHNF	21438	2J4ADV	Leased from Sterling Airways Delivered 8/4/80. Last service flown 27/10/91. Sold to Northern Lakes Finance.
G-BHVT	21349	212ADV	Obtained 3/6/80 ex-9V-SGC from Singapore Airlines Last service flown 25/10/82. Sold to LASCA, Costa Rica 28/10/82. Leased from LASCA 22/4/83. Returned off lease 31/10/83. Leased from LASCA 22/4/84. To Lasham 28/10/92 for disposal.
G-BIUR	19619	155C	Obtained 1/5/81 on lease from Ariana Afganistan. Last service flown 30/10/81. Returned off lease 25/11/81. Leased from Ariana Afganistan 22/4/82. Returned off lease 21/10/82.
G-BKAG	21055	217ADV	Delivered 1/4/82. ex C-GCPA of CP Air. Leased to Sun Country Airlines
G-BKCG	20328	170	Obtained 1/5/82 on lease from SAN, Equador. Last service flown 5/11/84. Returned off lease 18/12/84.
G-BMLP	20710	264ADV	Obtained 6/86 ex-N728ZV from Aeron Corp. Delivered 4/4/86. To Lasham 2/11/92 for disposal.
G-BNNI	20950	276ADV	Obtained 9/12/86 from Aeron Corp, (ex-VT-TBK of TAA). To Lasham 2/11/92 for disposal.
G-BPNS	20550	277ADV	Obtained from Aeron Coporation 17/1/89 ex-N276WC. To Lasham 2/11/92 for disposal.
G-BPND	21021	2D3ADV	Delivered 5/1/88. ex- PH-AHZ. To Lasham 2/11/92 for disposal.

*All down and out! 727 G-BFGN at Lasham.
(DAS/GMS Collection)*

DAN-AIR SERVICES

*Waiting for a flight to Athens. 727 G-BDAN rests at Gatwick.
(:DAS/GMS Collection)*

G-BPNY	20675	230ADV	Obtained from Condor 16/3/89 ex D-ABMI. To Lasham 3/11/92 for disposal.
G-NROA	21056	217ADV	Obtained from CP Air. Delivered 1/2/83 previously registered G-BKNG.

Boeing 737-200 series

G-BCIV	21528	2L9	Obtained 22/10/80.
G-BICV	21538	269	Leased from Guiness Peat Aviation, ex OY-API. Delivered to Gatwick 4/11/80. Last service flown 27/10/83. Returned off lease 11/83.
G-ILFC	22161	2U4	Delivered 1/11/83 on lease from International Leasing & Finance Corp, ex G-BOSL. Re-registered G-WGEL
G-BJXJ	22657	219	Obtained 4/3/82 ex N851L on lease from International Leasing & Finance Corp. Last service flown 28/10/87. Returned off lease 3/11/87.
G-BJXL	22054	2T4	Obtained on lease from International Leasing & Finance Corp. 11/11/83. Leased to Nordair 1985. Last service flown 27/10/86. Returned off lease 31/10/86.

*Boeing 737 series 200 G-BICV.
(DAS/GMS Collection)*

238

- THE AIRCRAFT

G-BKAP	21685	2L9	Obtained 27/4/82 on lease from Maersk Air. Last service flown 28/10/83. Returned off lease 11/83
G-BKNH	21820	210ADV	Delivered 30/3/83 ex 4X-BAA from Arkia, Isreal. Named *'Highland Prospect'*. To Gatwick 6/11/92 for disposal.
G-BLDE	22876	2E7ADV	Obtained 22/12/83 ex 4X-BAC from Arkia, Isreal. To Gatwick 5/11/92 for disposal.
G-BMDF	22875	2E7ADV	Delivered 26/3/84 ex-4X-BAB Obtained on lease from Arkia & operated in Arkia colours but with Dan-Air titles for a time - later flown in full colours. To Gatwick 2/11/92 for disposal.
G-WGEL	22161	2U4	See G-ILFC. To Gatwick 6/11/92 for disposal.

Boeing 737-300 series

G-BNNJ	24068	3Q8	Delivered on lease from International Leasing & Finance Corp. 12/2/88. To British Airways (Gatwick) Ltd.
G-BOWR	23401	3Q8	Obtained on lease from International Leasing & Finance Corp. 27/2/92 ex- OO-ILF. To British Airways (Gatwick) Ltd.
G-BUHI		300	Delivery expected February 1993
G-BUHJ		300	Delivery expected 1993
G-SCUH	23254	3Q8	Obtained on lease from International Leasing & Finance Corp. 4/5/85. To British Airways (Gatwick) Ltd.

Boeing 737-400 series

G-BNNK	24069	4Q8	Del.on lease from International Leasing & Finance Corp. 30/11/88. To British Airways (Gatwick) Ltd.
G-BNNL	24070	4Q8	Delivered on lease from International Leasing & Finance Corp. 11/1/89. To British Airways (Gatwick) Ltd.
G-BPNZ	24332	4Q8	Obtained on lease from International Leasing & Finance Corp. 7/6/90. To British Airways (Gatwick) Ltd.
G-BSNV	25168	4Q8	Obtained on lease from International Leasing & Finance Corp. 5/2/92. To British Airways (Gatwick) Ltd.
G-BSNW	25169	4Q8	Obtained on lease from International Leasing & Finance Corp. 12/3/92. To British Airways (Gatwick) Ltd.
G-BUHK	-	400	Delivery expected March 1993
G-BUHL	-	400	Delivery expected June 1993
G-BVNM	24163	4S3	Obtained on lease from Citycorp (ex G-BPKA Air Europe) 1991. To British Airways (Gatwick) Ltd.

Boeing 737 series 300 G-SCUH lifts off from Gatwick.
(:DAS/GMS Collection)

239

DAN-AIR SERVICES

G-BVNN	24164	4S3	Obtained on lease from Citycorp (ex- G-BPKB Air Europe) 1991. To British Airways (Gatwick) Ltd.
G-BVNO	24167	4S3	Obtained on lease from Citycorp (ex- G-BPKE Air Europe) 1991. To British Airways (Gatwick) Ltd.
G-TREN	24796	4S3	Obtained on lease from International Leasing & Finance Corp. 5/4/91 (ex -G-BRKG Air Europe). To British Airways (Gatwick) Ltd.

BAe 146

G-SCHH	E1005	100	Leased from British Aerospace, ex-ZD696 29/7/84. Later purchased. Last service flown 21/8/90. Sold to British Aerospace.
G-BRJS	E1004	100	Leased from British Aerospace, ex-ZD695 20/4/85. Last service flown 4/11/85. Returned to British Aerospace 12/11/85. Leased from British Aerospace 22/5/87. Returned off lease 17/6/87.
G-BKMN	E1006	100	Delivered 23/5/83. Previously registered G-ODAN. To British Aerospace July 1992
G-BKHT	E1007	100	Delivered 18/6/83. To British Aerospace July 1992
G-BPNP	E1002	100	Obtained on lease from British Aerospace 30/8/87. Returned off lease 27/11/87.
G-BPNT	E3126	300	Obtained 14/4/89 from British Aerospace. To Filton 6/11/92 for disposal.
G-BPNU	E3155	300	Obtained 16/8/90 from British Aerospace. To To Filton 6/11/92 for disposal.
G-BSRV	E	200	Obtained 26/7/91 on lease from British Aerospace. Returned from lease 29/7/91
G-BUHB	E	300	Delivered June 1992. To Hatfield 6/11/92 for disposal.
G-BUHC	E	300	Delivered June 1992. To Filton 6/11/92 for disposal.
G-BUHD	E	300	Delivery expected June 1993
G-BUHE	E	300	Delivery expected July 1993
G-BUHF	E	300	Delivery expected August 1993
G-BUHG	E	300	Delivery expected September 1993.

BAe 146 G-BHKT flies low over the city of London and the Palace of Westminster. (DAS/GMS Collection)

DAN-AIR SERVICES

Appendix Three
Key Dates

Throughout the 40 year history of Dan-Air Services Ltd there has been a tremendous amount of important events, both for the airline in particular and the industry as a whole. The following lists as many as are known in chronological order.

1953   Aircraft in service - 1 DC-3
21 May   Dan-Air Services Ltd registered as subsidiary of Davies and Newman Ltd with captial of £5,000.
16 June   Fleet consisted of a single Douglas DC-3, inherited from Meredith Air Transport. C. of A. was transferred this date. Chief Pilot was F. R Garside.
??   The first service was an ad hoc charter from the airlines base at Southend via Manchester to Shannon. The DC-3 was used on the second Berlin airlift caused by the Soviet blockade.

1954   Aircraft in service - 2 DC-3.
Three Avro Yorks acquired.

1955   Aircraft in service - 2 DC-3, 1 York.
The airline moved its base from Southend to Blackbushe and Dan-Air Engineering was established at Lasham to look after a number of Avro Yorks acquired from the RAF. IT charters started from Blackbushe.

1956   Aircraft in service - 2 DC-3, 4 York.
?? May   First scheduled service, from Blackbushe to Jersey started.
16 Nov.   Over 350 Hungarian refugees were airlifted from Austria to the UK. Completed 14/12/56

1957   Aircraft in service - 2 DC-3, 4 York. 1 Freighter, 1 Heron.
The first of the Bristol Freighters and a DH Heron acquired.

1958   Aircraft in service - 2 DC-3, 6 York, 2 Freighter, 1 Heron.

1959   Aircraft in service - 2 DC-3, 4 York, 3 Freighter, 3 Ambassador.
An Air Ministry contract to carry 'Black Knight' rockets to Woomera in Australia was obtained. Also BEA freight contract for London Heathrow-Manchester-Glasgow (Renfrew) service. IT charters started from Manchester. Three Airspeed Ambassadors obtained.

1960   Aircraft in service - 2 DC-3, 4 York, 3 Freighter, 2 Dove, 3 Ambassador. Carried 100,000 pax.
BEA awarded Dan-Air freight contracts from London Heathrow to Milan, Rome and Brussels. A pair of DH Doves acquired to start a network of scheduled services from Bristol and Cardiff.
4 April   Bristol-Cardiff-Liverpool route started.
1 May   Bristol-Cardiff-Isle of Man route started.
31 May   Blackbushe closed and the airline moved to Gatwick.
18 June   First schedule from Gatwick started - the relocated Jersey service.
16 July   Bristol-Cardiff-Basle route started - Dan-Air's first international schedule. IT charters were started from Gatwick to Holland, Belgium, France Germany and Spain, and trooping contracts were obtained.

1961   Aircraft in service - 3 DC-3, 4 York, 3 Freighter, 2 Dove 3 Ambassador.
4 Jan.   Plymouth-Cardiff/Bristol-Liverpool schedule was extended to Newcastle. Routes and Aircraft of Scottish Airlines (Preswick) obtained.
27 May   Prestwick-Isle of Man schedule started. The Isle of Man also served from

DAN-AIR SERVICES

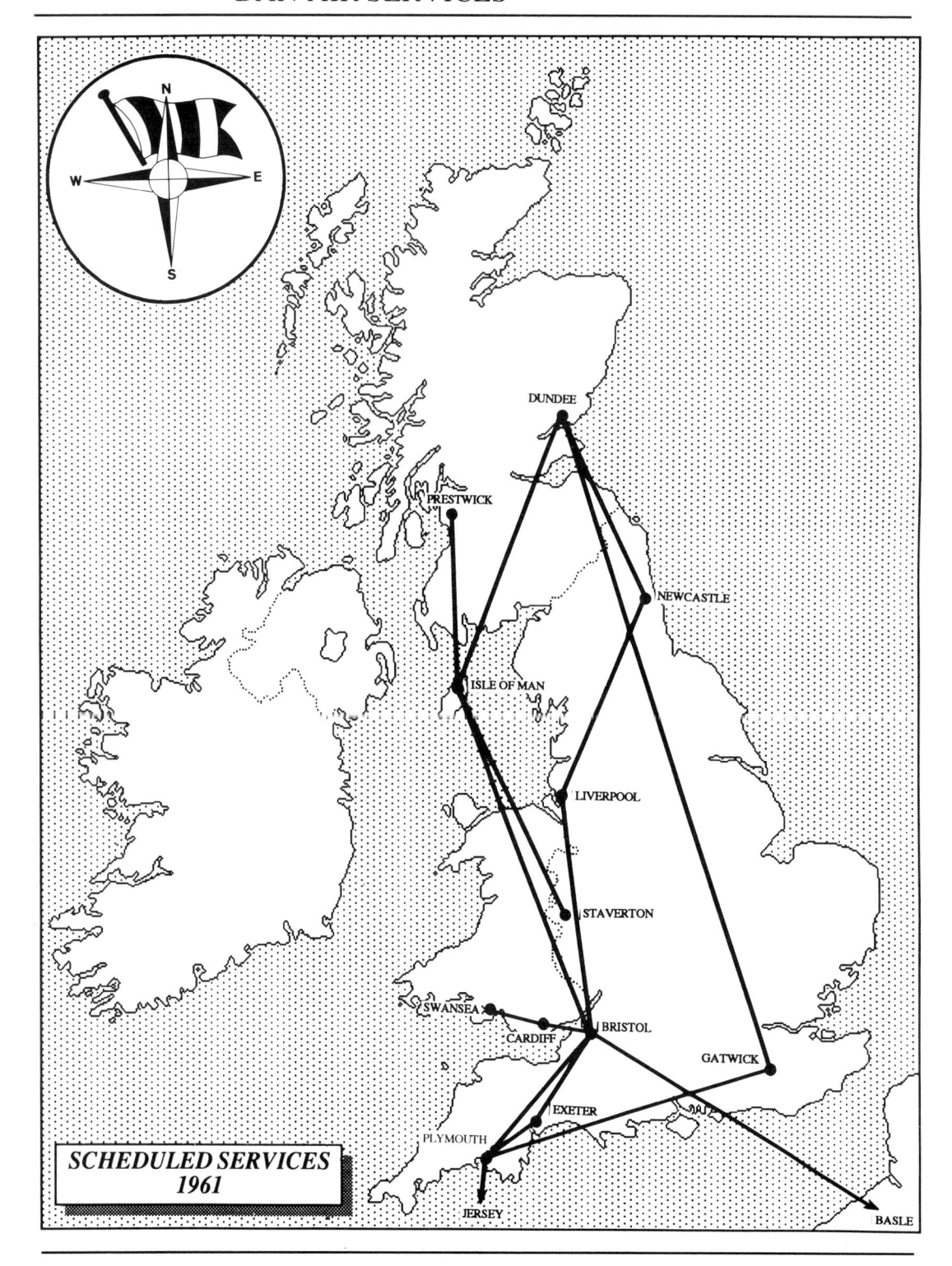

SCHEDULED SERVICES 1961

KEY DATES

A pair of Dan-Air cabin staff pose in front of a Comet. This picture must have been taken in the late 1960s for the aircraft is in the early, white tail colour scheme. Note the delightpul pillbox hats, white gloves and those hairstyles! (DAS/GMS Collection).

	Bristol, Cardiff, Plymouth, Exeter, Staverton and Swansea.
7 July	Liverpool-Newcastle-Dundee (Arbroath) route started, followed by Perth (Scone)-Prestwick-Gatwick and Perth-Newcastle-Gatwick that summer.
1962	Aircraft in service - 3 DC-3, 4 York, 3 Freighter, 2 Dove, 4 Ambassador. Second international service started: Liverpool-Rotterdam. Routes from Bournemouth to Basle and Bristol and Gatwick-Ostend began.
1963	Aircraft in service - 3 DC-3, 2 York, 3 Freighter, 1 Heron, 2 Dove 6 Ambassador. The Yorks were retired and a larger fleet of Ambassadors now formed Dan-Air's front line fleet for IT's and charters - mostly flying out of Gatwick and Manchester. Total pax for the year 115,215 (35,735 on schedules).
1964	Aircraft in service - 4 DC-3, 1 Avro York, 2 Freighter, 1 Heron, 1 Dove. 7 Ambassador.
1965	Aircraft in service - 3 DC-3, 2 Freighter, 1 Heron, 1 Dove, 7 Ambassador, 2 DC-4, DC-7. Tees-side-Chester-Cardiff service started, followed by a Cardiff-Bristol-Amsterdam and Gatwick-Newcastle-Kristiansand (Norway) services.
4 Oct.	Liverpool-Amsterdam service began. Two DC-4s and a DC-7 all-cargo Aircraft acquired.
1966	Aircraft in service - 4 DC-3, 2 Freighter, 8 Ambassador, 1 DC-7, 2 Comet. Obtained 2 DH Comet 4 airliners from BOAC - its first pure-jet aircraft and the beginnings of an ultimate 49 examples of the type (more than owned by any other operator).
1967	Aircraft in service - 3 DC-3, 2 Freighter, 6 Ambassador, 1 DC-7, 4 Comet. Comets went into service on Mediterranean IT charters. A Carlisle-Isle of Man route started in May. Two more Comets bought.
1968	Aircraft in service - 3 DC-3, 2 Freighter, 4 Ambassador, 1 DC-7, 4 Comet.

DAN-AIR SERVICES

1969   Aircraft in service - 1 DC-3, 1 Freighter, 3 Ambassador, 1 DC-7, 11 Comet, 3 1-11. Total pax carried 509,025.
Directors: Chairman - F.E.F. Newman, Managing - A.J.Snudden, Deputy Managing, R.A. Pigeon, Financial - E.M. O'Regan, Sales - F. Tapling, Ops Manager K. Balsdon. Chief Pilot, Capt R Atkins. With the demise of British Eagle in November 1968 Dan-Air picked up many more IT contracts and two ex-American Airlines BAC 1-11s were obtained and two more Comets to fly the extended programme. A Gatwick-Newquay service was started. The first trans-Atlantic flight (a Comet 4) to Trinidad was operated. First aircraft based in West Berlin to fly charters to the Mediterranean.

1970   Aircraft in service - 1 DC-3, 1 Freighter, 3 Ambassador, 12 Comet, 4 1-11, 1 262. Carried 1,000,000 pax.
Directors: Chairman - F.E.F. Newman, Managing - A.J.Snudden, Deputy Managing, R.A. Pigeon, Financial - E.M. O'Regan, Sales - F. Tapling, Ops Manager K. Balsdon. Chief Pilot, Capt R Atkins.
A Nord 262 obtained to replace DC-3s on schedules. Fleet also included 11 Comets 4s and 4 BAC 1-11s. The main IT programme was out of Gatwick, Luton, Manchester, Glasgow, Edinburgh and Tees-side.

4 July   Comet 4 G-APDN en-route Manchester-Barcelona crashed in the mountains at Arbucias, near Barcelona, killing all 105 pax and 7 crew.

29 Oct.   Granted a USA charter permit for trans-Atlantic charters.

1971   Aircraft in service - 1 Ambassador, 14 Comet, 5 1-11, 1 262, 1 707, 2 748. Carried 1,129,000 pax.

7 Jan.   Boeing 707-320 was obtained from Pan Am to start the Company's first trans-Atlantic charters in April.

1 May   The airline obtained its first HS748 airliner.
The airline mounted a big new charter programme out of Brimingham during

Rosemary Murch stands at the top of the steps of DC-3 G-AMPP. This is the famous mini-skirt era as can be clearly seen! (R Cooper Collection).

Passengers check-in for charter flights at Gatwick. The rust coloured uniforms denote Gatwick Handling staff (GMS/DAS Collection).

244

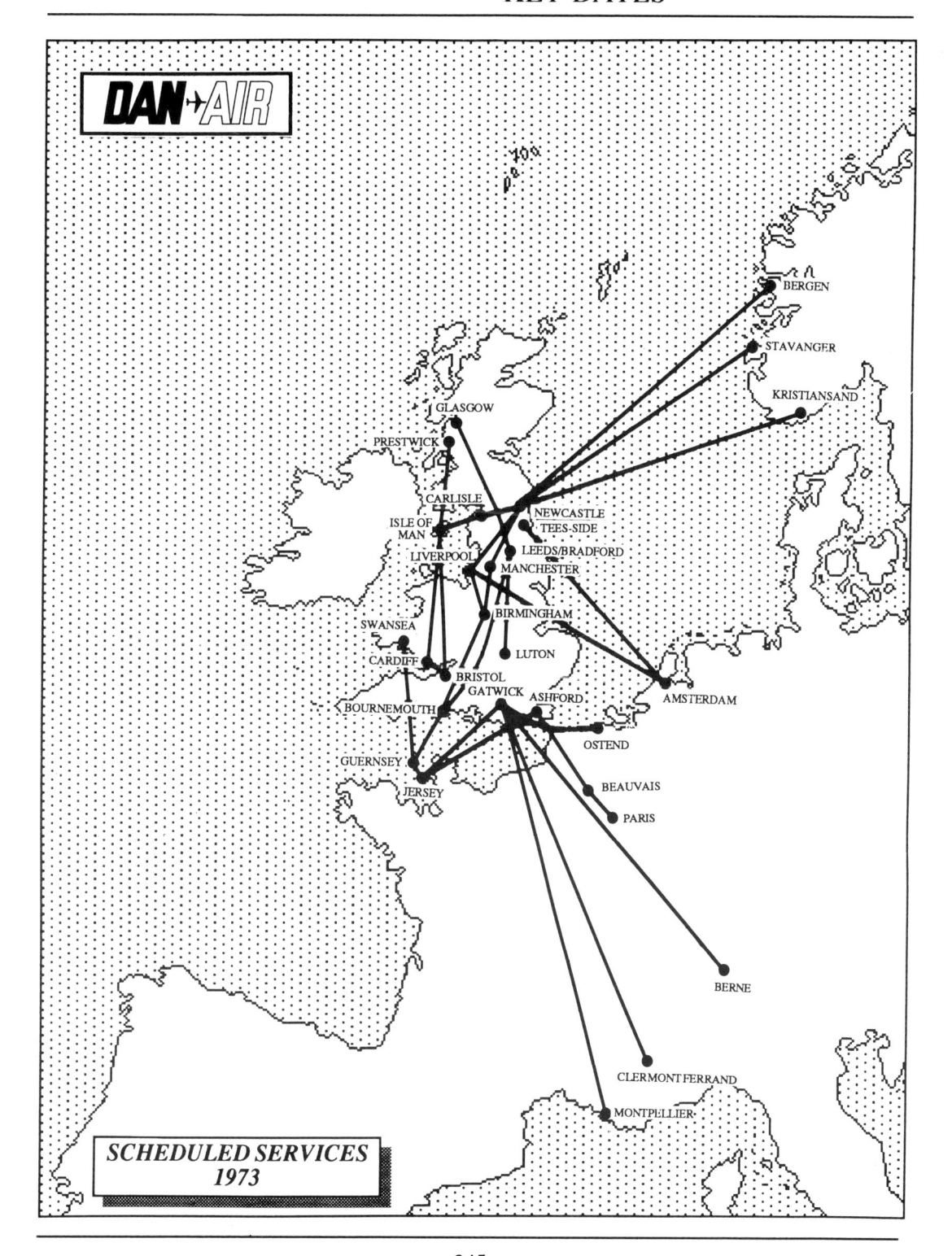

KEY DATES

DAN-AIR SERVICES

	the summer and also at Bristol and Cardiff, flying for all the major British tour operators.
26 Sept.	The last scheduled Ambassador service flown and type was phased out.
1972	Aircraft in service - 19 Comet, 5 1-11, 1 262, 2 707, 6 748. Carried 1,741,000 pax.
11 Feb.	The airline bought out the Ashford-based Skyways International who operated 4 HS748s. Another 748 was obtained from BOAC. Routes to Beauvais, Montpellier and Clemont Ferrand were inherited from Skyways and transferred to Gatwick.
10 April	Bournemouth - Birmingham - Liverpool/Manchester - Newcastle service started.
11 April	Luton-Leeds-Glasgow service started.
27 May	Swansea-Jersey and Newcastle-Carlisle-Jersey service started.
5 June	Gatwick-Berne service started.
1 July	Bournemouth-Guernsey/Jersey service started. A second Boeing 707-320 was obtained from Pan Am and several Comet 4bs from Channel Airways and BEA Airtours. Trans-Atlantic charters were flown from Gatwick, Manchester and Prestwick.
1973	Aircraft in service - 22 Comet, 5 1-11, 2 707, 7 748, 3 727. Carried 2,157,000 pax. Directors: Chairman - F.E.F. Newman, Managing - A.J.Snudden, Deputy Managing, F. Horridge, Financial - E.M. O'Regan, Sales - F. Tapling, Ops Manager E. Bristow, Chief Pilot - Capt. R Atkins. Three Boeing 727s were obtained from Japan Airlines, for use on IT charters from the UK and West Berlin, with Dan-Air becoming the first, and only operator of the type within the UK.
?? March	Tees-side added to the 'Link City' network.
2 April	Tees-side-Amsterdam service started.
May	Ashford (Lympne)-Jersey services started. After the worldwide fuel crisis late in the year the airline placed greater emphasis on 727/1-11 operations rather than the 'thirsty' Comets.
Winter	First located use of 'aircraft' as hypen in Dan-Air name; used on winter 73/74 timetable.
1974	Aircraft in service - 14 Comet, 5 1-11, 1 C-150, 2 707, 7 748, 5 727. Carried 2,193,000 pax. Directors: Chairman - F.E.F. Newman, Managing - A.J.Snudden, Deputy Managing, F. Horridge, Financial - E.M. O'Regan, Sales - F. Tapling, Ops Manager E. Bristow, Chief Pilot - Capt. R Atkins. The airline based its first HS748 in Aberdeen, for oil industry movements to Shetland and Orkney.
?? March	CAA granted the airline the first IT-affinity group charter licence ever for use on the Gatwick-Hong Kong flights.
14 April	A Newcastle-Isle of Man weekend service started and the Cardiff-Bristol-Amsterdam service re-started using 748 after several years suspension.
29 April	Twice daily Gatwick-Newcastle shedule was started using the Comets.
7 Oct.	A Leeds/Bradford-Bournemouth service was started and the Leeds-Luton service discontinued.
31 Oct.	Moved its cross-Channel services from Lympne to Lydd.
1975	Aircraft in service - 19 Comet, 13 1-11, 3 Viscount, 2 707, 9 748, 5 727. Carried 2,582,000 pax. Directors: Chairman - F.E.F. Newman, Managing - A.J.Snudden, Deputy Managing, F. Horridge, Financial - E.M. O'Regan, Sales - F. Tapling, Ops Manager E. Bristow, Chief Pilot - Capt. R Atkins

*Penny Stevens wears the dark blue red uniform with red piping that was introduced in 1975. Many thought that this outfit was supremely elegant..
(DAS/GMS Collection).*

KEY DATES

Thought to be another publicity photograph to promote the airlines meal service using 'Sky Diners' that this cabin staff member is serving to the few passengers aboard this Comet. (DAS/GMS Collection).

	Four ex-Court Line 1-11 500 series and two ex-Zambia 1-11 200 series machines were obtained to replace Comets.
16 March	The Liverpool-Amsterdam service was stopped and Liverpool dropped from the 'Link City' network.
17 May	1-11s started Gatwick-Ostend services.
18 May	1-11s started Gatwick-Jersey services.
20 May	1-11s started Newcastle-Bergen services.
	Newcastle-Kristiansand went to a full weekday service.
24 May	Tees-side-Isle of Man flights began, as did Aberdeen-Isle of Man and Gatwick-Isle of Man.
13 May	Gatwick-Berne operated daily.
1 June	Perpignan was added to the Gatwick-Clemont Ferrand route and a Gatwick-Belfast night freight service started.
	Boeing 707s operated ABC charter flights to the USA from Gatwick, Manchester and Preswick.
1976	Aircraft in service - 18 Comet, 14 1-11, 4 707, 11 748, 6 727. Carried 2,846,000 pax.
	Directors: Chairman - F.E.F. Newman, Managing - A.J.Snudden, Deputy Managing, F. Horridge, Financial - E.M. O'Regan, Sales - F. Tapling, Ops Manager E. Bristow, Chief Pilot - Capt. R Atkins.
6 Jan.	Bristol-Cardiff-Leeds-Glasgow service started with 748s.
13 Jan.	Announced that the Manchester Engineering base would be soon open.
26 Jan.	The airline used Television for the first time to sell its schedules, with adverts running on Southern TV.
17 April	Bournemouth-Dinard service started.
18 April	Carlisle-Jersey service started.
28 May	Newcastle-Stavanger service started.
20 June	Gatwick-Perpignan and Kristiansand service started.
	Charters were flown to Isreal for the first time, and MoD contract to Gibraltar was obtained.
1977	Aircraft in service - 17 Comet, 14 1-11, 6 707, 17 748, 8 727. Carried 3,591,000 pax.
	Directors: Chairman - F.E.F. Newman, Managing - A.J.Snudden, Deputy Managing, F. Horridge, Financial - E.M. O'Regan, Sales - F. Tapling, Ops Manager E. Bristow, Chief Pilot - Capt. R Atkins.
1 April	With the importance of the EEC, a Gatwick-Strasburg service was started.
4 April	A new Irish Sea route, Bristol-Cardiff-Cork commenced.
April	One Boeing 707 was leased to Air Malta.
14 May	Boeing 707-321C G-BEBP crashed on approach to Lusaka International Airport, Zambia at 0933 hrs GMT whilst on charter to I.A.S. Six crew lost.
21 May	East Midlands, Birmingham and Isle of Man triangular service began.

DAN-AIR SERVICES

The crew try their best to pose for a picture as the passengers scramble aboard G-BDIW for the last-ever commercial Comet flight on 9 November 1980. (John Hunt Collection).

1978   Aircraft in service - 13 Comet, 14 1-11, 5 707, 19 748, 8 727. Carried 4,010,000 pax.
1 April   The Silver Jubilee year saw services from Gatwick-Bergen commenced.
15 April   Gatwick-Jersey commenced.
27 May   Bournemouth-Isle of Man and and Jersey-Cork commenced.
     China's Vice-Premier Wang Chen chartered a 1-11 for his tour of the UK. Two B727-100 and a further HS748 were obtained.

1979   Aircraft in service - 7 Comet, 15 1-11, 2 Viscount, 1 707. 21 748, 8 727. Carried 3,548,000 pax.
Directors: Chairman - F.E.F. Newman, Managing - A.J.Snudden, Deputy Managing, F. Horridge, Financial - W. Jones, Sales - F. Tapling, Ops Manager B.V.S. Williams Chief Pilot - Capt. A Larkman.
Comet 4B fleet ended their tour of passenger service, the final aircraft being presented to the Science Museum.
1 April   Dan-Air and Intra Airways made each other joint licenses on Carlisle, Swansea and Staverton to Jersey/Guernsey routes, to be able to use each others aircraft.
1 April   Bristol/Cardiff-Guernsey route started.
23 May   Newcastle-Birmingham-Isle of Man route started.
1 May   Gatwick-Dijon route started.
31 July   At 1601 hrs 748 G-BEKF crashed on take-off from Sumburgh's 09/27 runway in the Shetlands. Of the 44 pax and 3 crew aboard, 15 pax and 2 crew (both pilots) lost their lives, but 29 pax and the air stewardess survived, the Stewardess later decorated for bravery in swimming in sea to rescue pax.
1 Nov.   Took over previously British Airways operated Gatwick-Aberdeen route.
16 Dec.   Gatwick-Toulouse route started.

1980   Aircraft in service - 4 Comet, 15 1-11, 2 Viscount, 20 748, 10 727, 2 737. Carried 3,510,000 pax,
28 Jan.   A custom-built HQ building (Newman House) was opened on 28/1 at Horley by Norman Tebbit MP to house Flight Operations staff - and later Accounts and Reservations.
Dan Air Engineeering obtain contract to look after the Boeing 737-200 fleet of Air Europe. Two Boeing 727-200s acquired in March and April for longer-range IT charters and substantial refurbishing of the 1-11 500 fleet carried out to give a 'wide bodied' look. Fin logo lights were added to the fleet.
19 Feb.   New package of low fares was announced on domestic routes, including 'Supersaver' excursion, 'Latesaver' last-minute and Spouse fares to start in April.

KEY DATES

Over the years Dan-Air undertook a number of charters carrying orchestras around. Here is the main cabin of a 727 converted to a mixed configuration, with the instruments of the Royal Philharmonic Orchestra under netting. The orchestra were on a major tour of Europe.
(John Coghlan)

1 April	British Airways Belfast and Dublin-Newcastle, Bristol and Cardiff; Bristol/Cardiff-Jersey, Guernsey, Paris and Leeds-Guernsey taken over.
25 April	Boeing 727-46 G-BDAN lost when destroyed after flying into Monte Chiriguel, Tenerife during the final stages of a charter flight from Manchester. 138 pax and 8 crew lost their lives.
1 May	Gatwick-Munich service started and rights granted to fly a West Berlin-Amsterdam service.
August	Reductions were sought in European fares and in September more domestic reductions weree introduced with 'Daysaver' fares.
26 Oct.	Winter schedules showed Gatwick-Aberdeen 3 times daily, and Newcastle had weekend flights. Llower fares announced for the winter.
9 Nov.	Last commercial Comet flight. The first B737-200 arrived to start service.

1981 Aircraft in service - 17 1-11, 1 Viscount, 21 748, 11 727, 2 737. Carried 3,226,000 pax.

16 Jan.	Cardiff/Bristol-Paris service ceased.
9 Feb.	The last Comet flight Gatwick-Dusseldorf took place.
1 April	West Berlin-Amsterdam and Tees-side-Dublin services started.
13 April	Gatwick-Cork service started.
	Summer schedules were started in March.

Boeing 727 G-BCDA at Stand One of Berlin-Tegel airport.
(GMS/DAS Collection).

DAN-AIR SERVICES

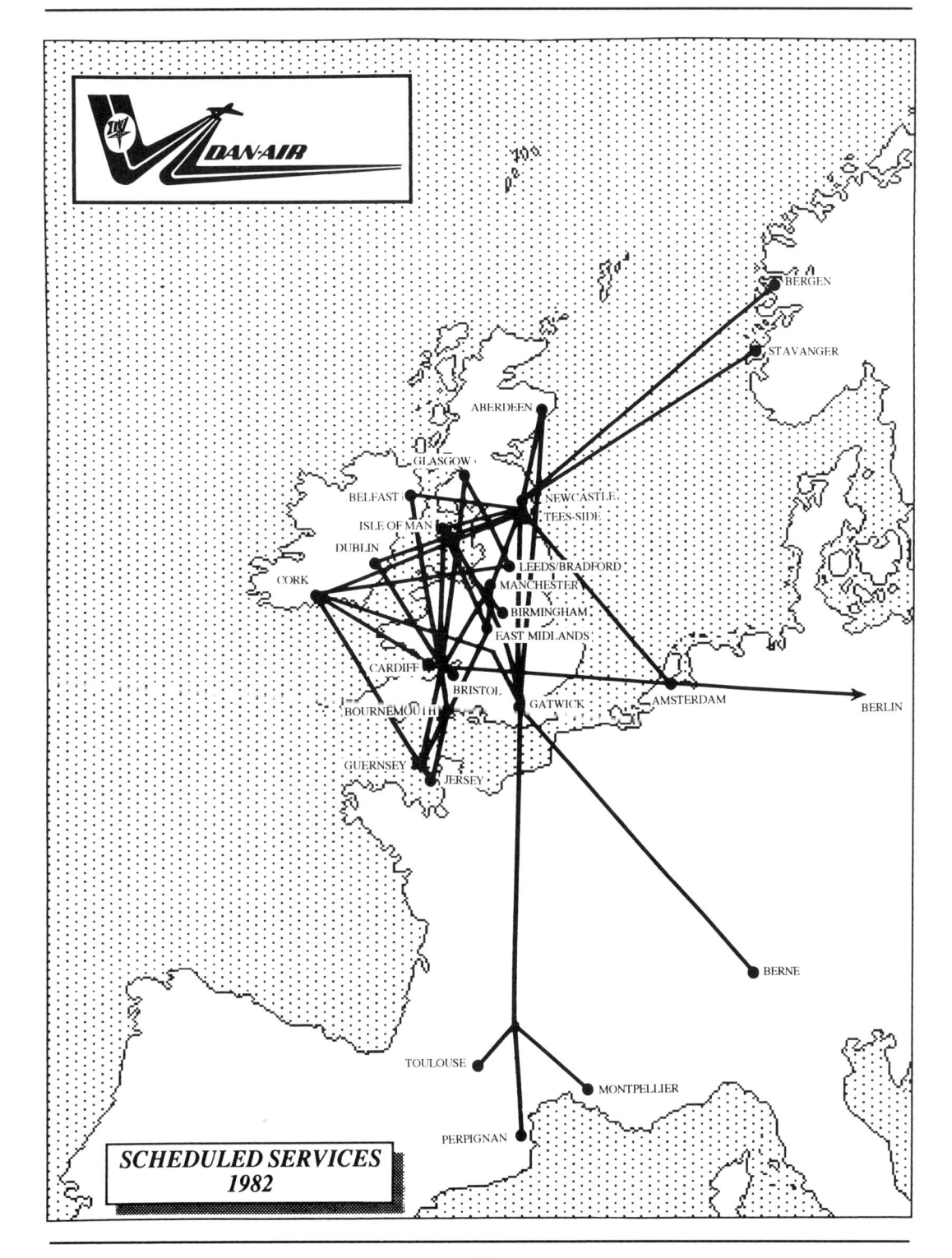

SCHEDULED SERVICES 1982

KEY DATES

9 May	Some 14 B727, BAC 1-11 and HS748 charters were operated into Shetland for the the official opening of the Sullom Voe Oil Terminal.
June	New Aberdeen-Newcastle and Newcastle-Jersey services started on the first weekend in June.
26 June	At 1811 hrs GMT 748 G-ASPL crashed at Nailstone Leicestershire whilst engaged on a mail flight from Gatwick to East Midlands Airport, with the loss of both pilots and the loadmaster.
27 Aug.	The CAA granted a Gatwick-Dublin licence to Dan-Air.
16 Nov.	The airline withdrew from its plan to take over the British Airways Highlands and Islands routes.
1982	Aircraft in service - 19 1-11, 19 748, 12 727, 4 737. Carried 3,599,000 pax.
26 Feb.	Three 748s leased to British Airways (Highland Division).
29 March	Metropolitan Airways, a commuter carrier, took over the 'Link City' network between Newcastle, Manchester, Birmingham, Cardiff and Bournemouth, operating under the Dan-Air banner.
March	A specially modified 748 with a large cargo door was put into service.
28 June	Daily Gatwick-Dublin service started.
11 Aug.	Reductions of 10% were of Norwegian fares and licence to operate Gatwick-Zurich granted.
16 Aug.	Dan-Air became the first airline to order the BAe 146-100 aircraft.
25 Oct.	New direct Bristol/Cardiff-Glasgow flight started using a Bandeirante.
28 Oct.	The airline applied to take over the British Airways London Heathrow-Inverness route when British Airways stopped in March 1983.
1983	Aircraft in service 20 1-11, 18 748, 11 727, 8 737, 2 146. Carried 3,702,000 pax.
27 March	Took over the London Heathrow-Inverness route from British Airways.
30 April	Gatwick-Zurich daily service started.
15 May	AGM of Davies & Newman PLC announce turnover of £196,127,000; Operating profit of £3,511,000.
21 May	Inverness Aberdeen and Gatwick services started.

Loading the night mail! Many of Dan-Air's aircraft carried the 'Royal Mail' emblem on the nose. Here 146 'NT is loaded with sacks. (DAS/GMS Collection).

251

DAN-AIR SERVICES

	The 146 took over services from Gatwick to Berne and Dublin, Perpignan and Toulouse and from Newcastle to Gatwick, Bergen, Stavanger and from Leeds to Jersey.
	Dan-Air became the first airline to advertise on London taxis.
23 May	Dan-Air became world launch customer for the BAe 146-100 when the first aircraft was delivered to Gatwick, a second following in June based at Newcastle.
1 Nov.	The airline signed to join the Travicom reservations system.
1984	Aircraft in service - 19 1-11, 1 Herald, 18 748, 11 727, 7 737, 3 146. Carried 4,559,000 pax.
9 Jan.	The airline started a Saarbrucken-West Berlin service, taking over the contract from a French airline.
5 Feb.	Took over the Gatwick-Belfast(Aldergrove) services from British Midland. Dan-Air operated twice-daily until 26/3 then 3 daily jet services and introduced the lowest fares on the route.
26 March	Gatwick-Zurich route went twice daily and the Berne services increased to nine weekly.
14 May	Daily Manchester-Zurich service started.
17 May	AGM of Davies & Newman PLC announce turnover of £242,846,000; operating profit of £1,841,000.
30 May	Gatwick-Jersey all year round service started.
16 July	Newcastle/Tees-side-Amsterdam service started.
July	Obtained a third BAe 146-100 (G-SSHH) that was used for starting the first jet services to Guernsey from Bristol and Cardiff).
	Leased three HS748s to Philippine Airlines for two years and 727 to Royal Air Nepal for six months.
	Gained approval to operate London Heathrow-Manchester in competition with British Airways Super Shuttle starting 1/4 1985. Also plans to start Inverness-Manchester on same date.
6 Aug.	Tees-side/Newcastle-Belfast service started replacing Genair.
	The Link City services between Glasgow, Leeds Cardiff and Bristol were handed over to Metropolitan Airways to operate.
	Dan-Air ordered its first Boeing 737-300.
	Obtained planning permission for a new hangar complex at Gatwick.
21 Dec.	Start weekly 146 Birmingham-Geneva for 16 week period of ski flights.

BAe 146 G-BPNT star of the weekend programme 'Airport 90' taxies in at Gatwick. (DAS/GMS Collection)

252

KEY DATES

1985	Aircraft in service - 20 1-11, 14 748, 9 727, 8 737, 4 146. Carried 5,007,000 pax.
?? Jan.	Applied for weekly Gatwick-Lourdes service.
1 April	Three times daily London Heathrow-Manchester service started. Manchester-Inverness and Manchester-Newcastle-Oslo service that connected at Newcastle with Bergen and Stavanger flights started.
6 April	Gatwick-Lourdes service started.
1 May	Resume Gatwick-Munich services 3 times a week after break of 4 years.
24 May	Manchester-Montpellier service started.
Sept.	London Heathrow-Manchester service discountinued. The EEC Lome Conference was hosted at Inverness with Dan-Air flying many extra services with delegates from London Heathrow.
2 Sept	Metropolitan Airways goes into liquidation following suspension of services on 31 August.
15 Dec.	Gatwick - Innsbruck service started. Metropolitan Airways planning to take over a number of domestic routes which it is currently operating under the 'Link City' network for Dan Air. Metropolitan had operating agreement with Dan-Air since 1982 when they provided the aircraft whist Dan-Air provided reservations and other facilities. Agreement was for this to happen for two years then Metro should take-over the routes. Apply to Civil Aviation Authority to serve Lisbon and Oporto from Manchester from 1987 and hopes to win appeal against rejection of Gatwick Lisbon service. Gain Civil Aviation Authority approval to operate London Gatwick-Santiago in Portugal and Seville subject to foreign government approval.

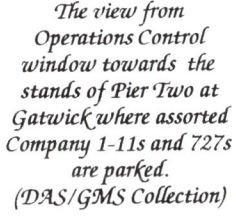

The view from Operations Control window towards the stands of Pier Two at Gatwick where assorted Company 1-11s and 727s are parked. (DAS/GMS Collection)

DAN-AIR SERVICES

1986   Aircraft in service - 19 1-11, 14 HS748, 10 727, 8 737, 2 Airbus, 3 146. Carried 5,309,000 pax.
Cut Gatwick-Aberdeen to twice daily due to down turn in Oil flights.
First known used of 'Aircraft' hypen symbol in Dan-Air name. Used on Airbus G-ABMNA.
31 March   Manchester-Amsterdam service started.
May   A300 proving flight with G-BMNA start to enter services.
Ends Bristol-Amsterdam services after 8 years.
Charters to Luxor started.

1987   Aircraft in service t - 18 1-11, 12 748, 10 727, 8 737, 1 Airbus, 5 146. Carried 5,481,000 pax.
6 March   First cabin staff course training male stewards completed.
4 April   Manchester-Lourdes service started.
4 May   Gatwick Lisbon service started.
Won approval for Gatwick-Toulon in Southern France
Operate Gatwick - Perpignan during winter for first time.
Dan-Air become members of IATA.

1988   Aircraft in service - 19 1-11, 11 748, 10 727, 10 737, 2 Airbus, 3 146. Carried 5,809,000 pax.
Second A300 Airbus acquired.
£10m hangar complex opened at Gatwick by HRH Princess Alexandra.
1 May   Gatwick-Madrid and Ibiza services started.
16 May   Gatwick-Mahon service started.
Refurbish 3 1-11s for use on 6 times a day Gatwick-Paris service due to start on 23 October.
23 Oct.   Gatwick-Nice and Manchester -Aberdeen services started.
Gatwick-Charles de Gaulle service started.
Gatwick-Manchester service started
1 Dec.   First Boeing 737-400 delivered.
Class 'Elite' business class introduced on Paris Charles de Gaulle and Nice service.

1989   Aircraft in service - 17 1-11, 10 748, 12 727, 11 737, 2 Airbus, 4 146. Carried 6,276,000 pax.
26/3/89   Increase weekly services Gatwick-Dublin from 9 to 12.
16/4/89   Announce withdrawal from Gatwick-Cork service.
17/12/89   Manchester-Berne 146 weekly service between 17 December and 18 March 1990 for use of winter sports visitors.
'Class Elite' introduced on Gatwick-Dublin, Toulouse, Madrid, Lisbon and Zurich services.

The famous Dan-Air Pipe Band. It was formed in late 1988 by James Banks, one-time Scots Guards Pipe-Major and household Piper to the Queen. The Band achieved Grade One status in just four years, which ably shows the depth of talent and musical ability there was in the Band and, apart from appearing in many Championships, they played a prominent part in many Dan-Air publicity appearances. Financial constraints meant that their last appearance for Dan-Air was on 14 June 1991 at Warwick Castle before the Princess Royal and the International Olympic Committee (DAS/GMS Collection)

254

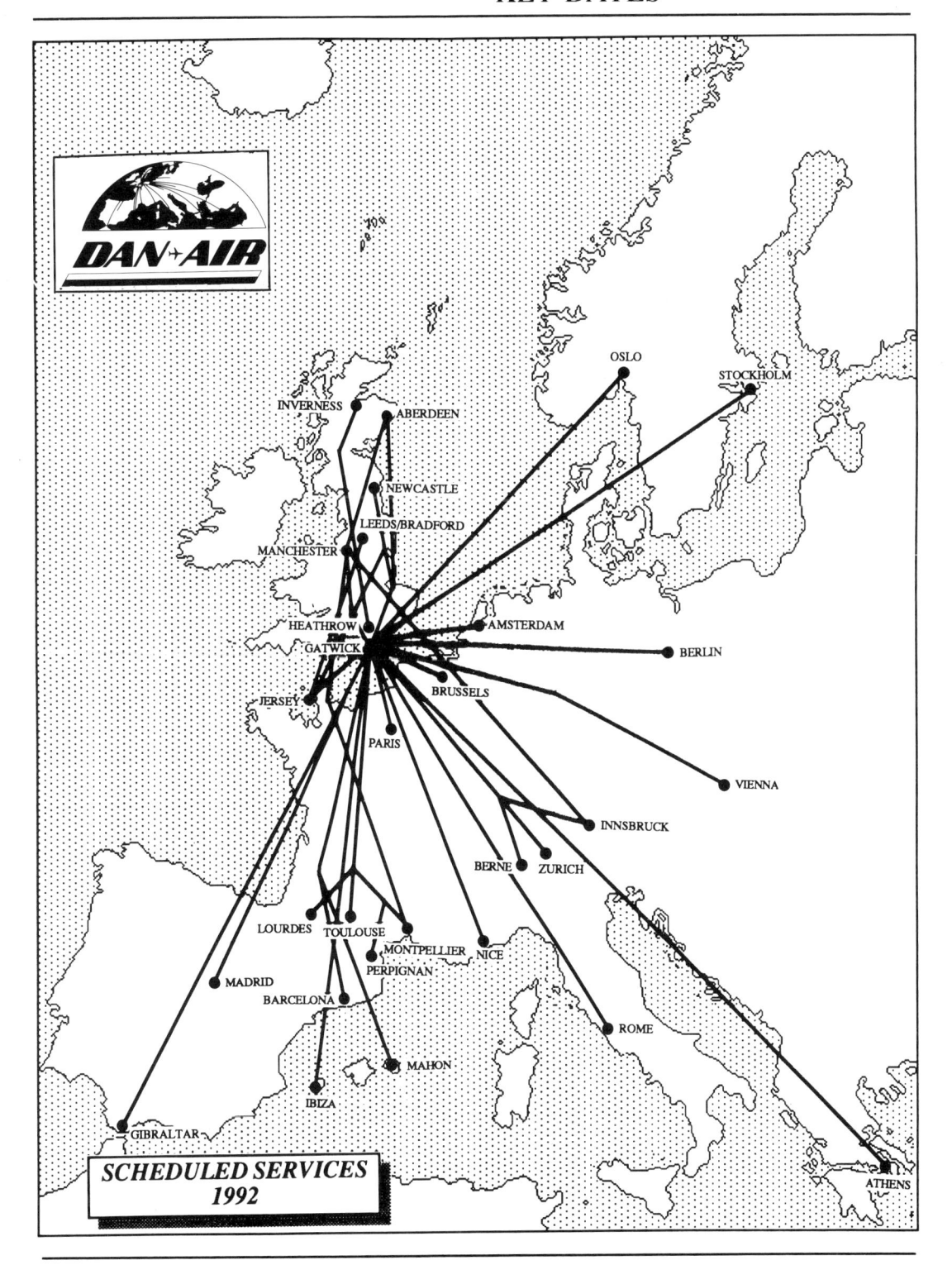

KEY DATES

SCHEDULED SERVICES 1992

DAN-AIR SERVICES

1990　　Aircraft in service - 18 1-11, 9 748, 10 727, 12 737, 1 Airbus, 5 146.
2/4/90　　Gatwick-Berlin service started.
4/90　　Gatwick-Dublin service discontinued.
2/5/90　　Gatwick-Vienna service started.
　　　　'Class Elite' introduced on Gatwick-Ibiza service
29/10/90　Through flights introduced between Manchester, Newcastle, Tees-side and West Berlin.

1991　　Aircraft in service - 15 1-11, 6 748, 7 727, 16 737, 5 146.
28/2/91　Completed sale of its aircraft engineering business to FLS Aerospace Ltd for £27.5 million
25/10/91　Sale was completed of Davies & Newman Ltd to a subsidiary of London Airtaxi Centre Ltd.
10/11/91　Twice daily except Saturday Gatwick - Oslo service started.

1992　　Aircraft in service - 12 1-11, 4 748, 7 727, 19 737, 4 146.
8/1/92　Completed sale of its shares in Manchester Handling Ltd to Gatwick Handling Ltd and Aer Lingus UK Holding Ltd.
26/2/92　Twice daily except Saturday Gatwick - Stockholm service started
2/3/92　Daily Gatwick - Ahtens service started.
5/4/92　Daily Gatwick Rome service started.
18/5/92　Daily except Saturday Gatwick - Barcelona service started.
2/10/92　Gatwick - Barcelona service closed.
23/10/92　British Airways and Dan-Air Announce that Dan-Air is to become a wholly owned subsidiary of British Airways PLC. The Dan-Air name will dissapear. Between 1400 and 1600 jobs will be lost and 26 of the 38 aircraft in the Dan-Air fleet disposed of.
8/11/92　At 12.19hrs the deal is completed transferring ownership to British Airways.

and finally... possibly the authors favorite picture, for it demonstrates just what the 'Spirit of Dan-Air' was all about. The date was 5 November 1992, the place Dalcross. Out of work but still smiling, the picture shows the Inverness-based staff grouped around the bronze statue of Captain E.E. Fresson, the pioneer of many of the Scottish airline routes. This photograph contains echos with the past, for Fresson's airline was nationalised into BEA in 1947 - Dan-Air Services amalgamated into their successor, BA, in 1992..

256